When No one's Coming

The Ultimate Scientific Guide to Staying Alive in a Nuclear War

Lucas Goldwire BSc(Hons), MPhil

When No One's Coming.

Copyright © 2023 by Lucas Goldwire.
All rights reserved.

Copyright Notice

No part of this book may be reproduced or transmitted by any means, except as permitted by UK / US copyright laws, or by the author. For licensing requests, please email requests@10secondstomidnight.co.uk

Photographs, pictures and diagrams are copyright © Lucas Goldwire, unless otherwise noted.

Disclaimer:

This publication is designed to provide accurate and authoritative information with regard to the subject matter covered. It is sold with the understanding that neither the author nor the publisher is engaged in rendering legal, investment, accounting or other professional services. While the publisher and author have used their best efforts in preparing this book, they make no representations or warranties with respect to the accuracy or completeness of the contents of this book and specifically disclaim any implied warranties of merchantability or fitness for a particular purpose. No warranty may be created or extended by sales representatives or written sales materials. The advice and strategies contained herein may not be suitable for your situation. You should consult with a professional when appropriate. Neither the publisher nor the author shall be liable for any injury, damage, or loss of profit or any other commercial damages, including but not limited to, special, incidental, consequential, personal, or other damages.

Cover Design by Liam Lacey
First Edition 2023 ISBN: 9798393699772

When No One's Coming.

When No One's Coming.

DEDICATION

This book is dedicated to the memory of my Father.

When No One's Coming.

CONTENTS

Acknowledgments

About the Author   1

1   Introduction   3

2   World Politics & the Doomsday Clock   9

3   Radiation, Threats, and Effects   15

4   Nuclear Risks   27

5   Hazards created by a Nuclear Emergency   39

6   Early Warning Systems and False Alarms   47

7   Personal Qualities Required for Survival   59

8   What to do in a Nuclear Emergency
      i)   Before the Event   63
      ii)   Immediate Action   79
      iii)   Short Term Action   83
      iv)   Medium Term Action   85

STRATEGIES FOR SURVIVAL

9   How to decontaminate   87

10   Monitoring Radiation Levels   93

11   First Aid   99

12	Protecting against Iodine-131 Exposure	131
13	Protection against Electromagnetic Pulse (EMP)	135
14	Expanding Your Emergency Supplies	141
15	Food	147
16	Food Storage and Shelf-life	155
17	Water	179
18	Other Aspects of Nuclear Survival	183
19	Conclusions	211
	Appendix I: Emergency Plan Template	217
	Appendix II: The Nuclear Emergency Canista$^{(TM)}$	225
	Appendix III: Useful phone apps	235
	Appendix IV: Glossary of Terms	237
	Appendix V: Bibliography	257

When No One's Coming.

TABLE OF FIGURES

Figure   Description

1   Castle Romeo Thermonuclear Test   7
    Detonation
2   Types of Radiation   16
3   Baker Nuclear Test Detonation   21
4   Nuclear Winter   25
5   Sources of Ionising Radiation   29
6   Abandoned Gas Masks at Pripyat, near   37
    Chernobyl
7   Emergency Siren   57
8   Nuclear Contamination Map of UK after   65
    Hypothetical Nuclear Strikes
9   Nuclear Contamination Map of USA after   67
    Hypothetical Nuclear Strikes
10 Map of UK showing Zones of Contamination 69
    after Potential Nuclear Reactor Accident
11 Where to go in nuclear emergency – FEMA 79
    guide diagram
12 How a Geiger Counter Works   93
13 Radiation Survey Meter in Operation   97

14 Tourniquets 101

15 Locations for Application of Tourniquets 103

16 A First Aider Practices CPR on a Mannequin 107

17 Suturing Kit and Practice pads 121

18 Suturing a Wound 125

19 Examples of Flexible Splint Applications 129

20 Food Storage in Reused PET Containers 173

21 Initial Stages of Improvised Nuclear Shelter Construction 185

22 Final Stages of Improvised Nuclear Shelter Construction 187

23 Radio Transceivers 197

24 Temperature reading from Evacuated Solar Tube 201

25 Evacuated Solar Tube Heating Water on an Overcast Day 201

26 Solar tubes used as kettles for boiling water 201

27 Evacuated Tube Solar Collector with Integrated Water Storage Tank 203

28 Roof-Mounted PV System 209

29 Gas Masks Awaiting Armageddon 215

Figures 8, 10, 14, 15, 17, 19, 23, 24, 25, 26 & 27 copyright the author.
Figures 1-7, 12, 13, 16, 18 & 28 licensed stock images.
Figure 9 & 11 FEMA.
Figure 20 by kind permission The Provident Prepper.
Figures 21 & 22 Crown Copyright, by kind permission of The National Archives Office.

When No One's Coming.

ACKNOWLEDGMENTS

I wish I had the opportunity to extend my heartfelt gratitude to my friend Martin, who sadly passed away several years ago. We spent many hours, late into the night, discussing climate change, off-grid living, renewables, energy dependency, the pros and cons of nuclear energy and nuclear bombs. Although I only ever met Martin maybe half a dozen times, we worked together online almost every evening for around 10years. I hope he would have approved of, and enjoyed, reading this book.

I also wish to thank Liam for producing the cover artwork – Let's hope we never witness it for real! Thank you to my proofreaders, who read and checked the manuscript for me, to the two supermodels who posed for the flexible splint photos, and to my family for putting up with my distraction during the writing of this book.

When No One's Coming.

ABOUT THE AUTHOR

Lucas Goldwire has written this manual from a scientific perspective, having obtained an Honours degree in science, and a Masters research degree in Medical Biochemistry. He spent several years working as a research scientist in a laboratory, including time spent working in a radiation laboratory. Goldwire has founded and run two multi-million pound businesses in the areas of Alternative Fuels and Renewable Energy as well as several small businesses reflecting his own interests in wide and varied fields– such as teaching motorcycle stunt riding (which was featured in a national motorcycle magazine) and identifying heat-loss from buildings, using thermal imaging - his most notable subject being Buckingham Palace in the UK – his images being featured on the front pages of several major UK newspapers at the time. Goldwire is currently a serving Firefighter and is trained in First Responder Emergency Medical Care and is a qualified Emergency Response Driver.

In this manual, Goldwire uses his expertise to cut through the usual myth and hearsay, with sound advice backed up by scientific principles, giving sensible, realistic suggestions and information essential to survival in a nuclear attack.

When No One's Coming

1 INTRODUCTION

Welcome to this comprehensive emergency manual, written from a scientific perspective to help you prepare for, and deal with, a nuclear emergency. The world we live in today is unfortunately a far less safe place than it was in the past. Ever since the invention of the nuclear bomb, the threat of nuclear war has loomed over humanity, and recent breakdowns in relations between nuclear powers, particularly since 2022, have made the world an even more dangerous place.

The possibility of a nuclear attack is now more real than ever before, and it is important to recognize the potential for rogue states and for desperate political leaders who may feel they have no option other than to launch a nuclear attack. While many believe that superpowers would never strike first and that they recognise no one can win a nuclear war, the fact is that the world is a complex place, and there is always the potential for miscalculation, misunderstanding, miscommunication and desperation Even a small mistake or accident could potentially trigger a catastrophic chain of events.

In the event of a nuclear emergency, if you need help, you can be sure that no one is coming. You are on your own and you will need to be equipped materially, mentally and physically to deal with it yourself. Emergency services will be stretched to the limit, and you will be forced to rely on your own resources to survive. Everything will change, and it will do so fast. You will need to react positively and decisively to this change. This manual aims to provide you with the knowledge and understanding that you will need to cope in the aftermath of a nuclear emergency, when you may be forced to rely solely on your own skills and knowledge to keep you and your loved ones alive.

The first step in preparing for a nuclear emergency is to understand what it is and how it works. Nuclear explosions can produce a wide range of harmful effects, including blast shock waves, thermal radiation, and ionizing radiation. Each of these effects can cause serious injury or death, and it is important to understand how to protect yourself and your loved ones from them. The aim of this book is to convey the scientific, and sometimes complex nature of atomic energy, in layperson's terms. If, when reading, you encounter a technical term which you do not understand, please refer to the glossary at the back of the book, which is included to avoid making the text of the book too laborious.

This manual will guide you through the steps you need to take to prepare for a nuclear emergency, including how to create an emergency plan, how to stockpile supplies, and how to protect yourself and your home. It will also provide you with information on the different types of radiation, how they affect the body, and how to minimize your exposure to them. Additionally, you will learn about the signs and symptoms of radiation sickness and how to treat it.

By following the guidance in this manual, you can develop your own plan and take action to achieve and maintain personal safety in the event of a nuclear emergency. You will also learn about the long-term effects of radiation exposure and how to minimize your risk of developing cancer or other illnesses in the years following a nuclear emergency.

In short, whether you are a concerned citizen, a prepper, or simply someone who wants to be better able to cope with a disaster, this manual is a valuable resource that can help you navigate the challenges of a nuclear emergency. The knowledge and tools provided can help you to survive and to minimize your exposure to the long-term effects of radiation. So, take the first step today, and arm yourself with the knowledge that will allow you to prepare for the worst, whilst hoping for the best.

When No One's Coming

Figure 1: *Beautiful but deadly – one of the most iconic nuclear bomb photographs, demonstrating the destructive power of the atom, unleashed in an enormous billowing mushroom cloud, the centre of which can reach, or even exceed a staggering 100 million °C – over five times hotter than the centre of the sun. This is the 1954 'Castle Romeo' thermonuclear test, conducted at Bikini Atoll, equivalent to 11Mt of TNT explosive.*

When No One's Coming

2 WORLD POLITICS & THE DOOMSDAY CLOCK

Most people will appreciate that there is a link between world politics and nuclear safety, and a generation of people has grown up under the threat of the Cold War – something that we all thought was behind us but has alarmingly reared its head again in recent times.

Nuclear History

The detonation of the nuclear bombs at Hiroshima and Nagasaki, in Japan in 1945, effectively ended the Second World War. It also initiated an arms race - stockpiling of nuclear weapons, firstly by the USA, followed in 1949 by the USSR (Russia) after successfully completing its first nuclear detonation in Kazakhstan. This continued until the end of the cold war in 1991, at which time, over 70,000 nuclear warheads existed. After the Cold War ended, both superpowers substantially cut back on their nuclear weapons stockpile. Even so, the total number of nuclear warheads worldwide today remains at around 13,000. Worryingly, some of the warheads are very much larger and more powerful than the ones used against Japan at the end of the Second World War. The largest nuclear bomb ever detonated was the Russian *Tsar Bomba,* also known as *Big Ivan*, which had a blast equivalent to 50 Megatons of TNT – over 3800 times more powerful than the bomb dropped on Hiroshima and would have caused third degree burns to observers at a distance of 62miles. There are also now 'tactical' nuclear warheads – small nuclear bombs for use on the battlefield, with destructive power ranging between less than 1 kiloton, up to about 100 kilotons (By comparison, the Hiroshima detonation was equivalent to 15 kilotons).

Currently there are 9 countries that possess nuclear weapons – the USA, Russia, the UK, France, China, India, Pakistan, Israel and North Korea. In addition to these, Iran is believed by many to be working hard to acquire its own nuclear weapons.

In the event of an aggressive nuclear detonation, it is likely that a retaliatory nuclear attack would ensue, in turn leading to further nuclear exchanges between adversaries, potentially escalating to a nuclear World War. These multiple strikes would likely destroy all countries involved and kill hundreds of millions of people. It would also result in massive nuclear contamination worldwide.

Country	No. of nuclear warheads
Russia	6257
USA	5550
China	350
France	290
United Kingdom	225
Pakistan	165
India	156
Israel	90
North Korea	50
Iran	Suspected development

Nuclear Winter

Climate modelling software has also shown that following a major exchange of nuclear weapons, global temperatures, particularly those on land would reduce significantly – in a hypothetical USA/Russia nuclear war, the nuclear winter that would follow, would see freezing temperatures during mid-summer for much of the Northern Hemisphere. Nuclear Winter is caused by the injection of huge amounts of smoke particles into to the stratosphere, originating from multiple fires and firestorms caused by the thermal radiation of the nuclear bombs. This soot would block out a significant percentage of solar radiation, resulting in crop failures, in turn leading to a further consequence – global famine, which could result in the deaths of ten times as many people as were killed by the nuclear explosions themselves. According to computer modelling, the stratospheric soot that causes nuclear winter could take a few years to clear. It is estimated that the Global Famine resulting from a global nuclear war would likely kill over 5 billion people.

Yet another effect would be the accelerated loss of the ozone layer, caused by stratospheric temperature increases of around 30°C for periods of years. In a regional nuclear war, this is likely to cause 25% destruction of the ozone layer, whereas in a global nuclear war, destruction is likely to be 75% of the ozone layer, with a recovery time likely to take decades, causing many more deaths due to skin cancer.

Scientists' Warning

Immediately after the two atomic explosions of the Second World War in 1945, the *Bulletin of the Atomic Scientists* was formed by former Manhattan Project members, as a non-profit organisation concerned with the potential negative impacts of some aspects of science and technology and has been continually active ever since. Although the organisation is not particularly well known, almost everyone will have heard of their most notable concept – the Doomsday Clock. This was conceived in 1947, as an indicator of how close humanity has come to creating a global catastrophe (as opposed to natural catastrophes) – from any type of threat - not just nuclear. However, since the proliferation of nuclear weapons, and the nuclear arms race of the 1980s, nuclear war has, not surprisingly, featured as the most likely contender, along with, more recently, Climate Change. Midnight, on the hypothetical Doomsday Clock, is considered the point at which a global catastrophe event takes place, potentially ending human civilisation.

At its initiation, the Doomsday Clock was set at 7minutes to midnight, indicating, symbolically and in the opinions of the scientists making up the *Bulletin of the Atomic Scientists,* that we were fairly close to creating a global catastrophe - in recognition of the start of the *Cold War.* The clock is recalculated each January, and has moved forwards and backwards, reflecting political stability of countries with nuclear arsenals, wars in which they are involved, terrorism and changes in threats to our climate. The furthest from midnight it has ever been, is 17 minutes to midnight, because of the end of the Cold War in 1991. The closest it has ever been set to midnight is 90 seconds to midnight – in January 2023 – a sobering thought. This is largely due to instabilities created by the Russian invasion of Ukraine.

At the time of writing, the biggest concern is that the Ukraine War could escalate into a nuclear conflict between Russia and NATO allies who are supporting Ukraine. However, there are also concerns that a nuclear conflict could develop between India and Pakistan or between the USA and North Korea, or possibly even between Iran and Israel in the Middle East. Unfortunately, owning nuclear weapons does not necessarily make you safer.

When No One's Coming

3 RADIATION, THREATS, AND EFFECTS

Before we can start planning how to minimise our exposure to radiation, we need to understand the different types of nuclear radiation and how they affect us. All types of nuclear radiation are known as *ionising radiation*. This means it has the capacity to strip electrons from atoms that it hits. This can have the effect of breaking chemical bonds, and if those bonds happen to be those that hold your DNA together, it can result in damage to the DNA, causing a mutation. In some cases, these mutations can lead to uncontrolled cell replication, which is basically a cancer. So ionising radiation has the potential to cause cancer. The more ionising radiation you are exposed to, the more likely it is to cause cancer.

The ionising – and therefore harmful – emissions caused by radioactivity can take a number of different forms:

Alpha Particles consist of helium nuclei (2 protons stuck together with 2 neutrons), which are ejected at high speed from certain radioactive elements when they decay. They are however, the most damaging to human tissue, but they are also the easiest to stop, as they can be absorbed very easily Even a thin sheet of paper is enough to halt alpha particles.

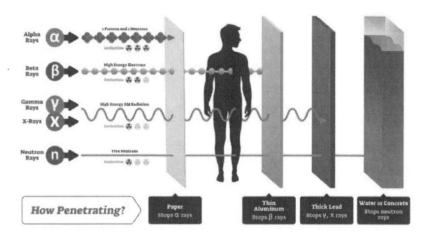

Figure 2: The ability of different types of radiation to penetrate various thicknesses of material.

Because of this, most detector instruments do not react to alpha particles, as they would be absorbed by the casing of the detector head itself. The outer layers of our skin, and our clothes, are very able to protect us from alpha-emitters, but the real danger comes when radioactive alpha-emitters are swallowed or breathed in – placing the highly damaging alpha particles inside the human body where it can do maximum damage to surrounding tissue.

Beta Particles are less damaging but much more penetrating than Alpha Particles. They are basically high-speed electrons ejected from some radioactive atoms as they decay. They can easily travel a couple of metres in air. They can penetrate deeper into the human body and can therefore cause more damage when presented externally. Beta particles can be easily shielded by aluminium foil, or a few centimetres thickness of plastic. Lower energy beta particles can be stopped simply by layers of clothing.

Gamma Rays are very high frequency, high energy electromagnetic waves (like light and heat but with much higher energy). Although gamma rays are less harmful than alpha or beta particles, they are much more penetrating, so can cause harm at a much greater distance, even with heavy shielding between the radiation source and the body. A few metres of concrete, or more than 30cm thickness of lead, is required to make a substantial reduction in gamma ray intensity.

In addition to the three 'classic' forms of radiation described above - that you would find in physics textbooks - there are other- hazards associated with nuclear emergencies:

X-rays are produced in a nuclear explosion and in nuclear reactors. Although these high energy rays are not as harmful as gamma rays, they are still harmful, and may be produced in high intensity during a nuclear incident, causing a significant health risk in its own right.

Neutrons are produced in both controlled and uncontrolled chain reactions present in nuclear power stations and in nuclear bombs. Both produce high quantities of neutrons, which are even more penetrating than gamma rays, and therefore not easy to shield against. Some types of nuclear bomb are specifically designed to produce very high levels of neutrons in order to kill a much greater number of humans than with a standard nuclear bomb. Neutrons have the potential to cause considerable damage to human tissue and when they penetrate the human body, they may bind to the nuclei of atoms inside your body, creating unstable, radioactive isotopes out of the very material that the body is made from. Neutron radiation can be considered up to 20x more harmful than gamma ray emissions. Concrete or water are the best shielding materials for neutrons, but for full protection, a few metres of shielding material is required. When the neutrons are absorbed by the shielding material, the process can produce secondary gamma rays, which can achieve harmful levels, so additional gamma ray shielding (such as lead) on the inside of the concrete or water shielding is also required.

Nuclear Fallout – Whether nuclear fallout occurs, and to what extent it occurs, depends on the height at which the nuclear bomb is detonated. There are four scenarios here:

1)A surface detonation (or an air detonation which is low enough that the fireball impacts with the ground). These are known as **'Surface Bursts'**. This type of detonation would be used where there is a desire to eliminate underground bunkers or silos that are well-protected against bombing. This type of detonation produces the largest amount of relatively large fallout particles – as the blast will draw up material from the ground – earth, houses, plant material etc, a fair amount of which will be vaporised and mixed with highly radioactive remnants of the nuclear fuel itself, the weapon casing and controls, together with radioactive transmuted isotopes from the surface material sucked up by the explosion, caused by absorption of the neutrons created by the blast. The resultant plasma will condense into large amounts of highly radioactive particles between 100nm and a few millimetres across, which will start dropping from the mushroom cloud as the blast subsides.

2) Detonations over water ('**Water Surface Bursts**') will produce substantial quantities of smaller, lighter fallout particles, which will take longer to settle out of the air, and therefore fall over a much larger area. Detonations over seawater will draw water up into the air, and the salt within this water may act as cloud-seeding, initiating rain shortly after the explosion has subsided. Such fallout is much more difficult to decontaminate, as the salts within the fallout will react with materials such as steel or concrete, causing them to chemically bond to the material. Chemically bonded fallout cannot be removed by washing, so material will need to be stripped away by processes such as sandblasting, in order to fully decontaminate.

3) Low altitude '**Airburst**' detonation (usually 100 - 1000m) – If the blast is high enough to avoid material from the ground interacting with the fireball, then much less fallout is created, but the fallout tends to be much more radioactive, and therefore more carcinogenic, as it is almost entirely made up of nuclear materials mixed with condensed remnants of the weapon casing and control gear. However, if it is closer to the ground, then the very high temperatures at the centre of the fireball causes the gases of the fireball to rise rapidly due to convection, in turn causing a vacuum that can drag debris upwards from ground level – this column of ascending material forms the stem of the characteristic 'mushroom cloud' of a nuclear explosion. If this stem meets the head of the mushroom i.e., the fireball, then the debris from the

ground will mix with the highly radioactive material, creating a much greater quantity of radioactive fallout.

Surface and low-altitude bursts may cause a more rapid fallout if rain forms soon after the explosion, as this will wash the dust out of the air much more quickly. In this situation, the rain falling will be very highly radioactive, and will immediately contaminate ground water, streams and rivers, as well as any land or buildings. As the rain causes fallout to descend a lot faster, a higher concentration is likely to be deposited in close proximity to the centre of the blast, as it will have had very little time to disperse over a wider area.

4)'**High Altitude Burst**' (above 30km) – these weapons are intended mainly to produce strong NEMPs (Nuclear Electromagnetic Pulses) to cause disruption, as the blast will have little or no effect on the surface.

Airbursts tend to generate a large proportion of very small radioactive fallout particles ranging from10nm - 20μm (10 nanometres to 20 micrometres, i.e., one ten thousandth of a millimetre to 0.02 millimetres), which are raised up into the stratosphere, where they remain for months or even years, during which time, they will be distributed worldwide by winds before finally falling to the ground – thus causing contamination of the entire world surface, although obviously diluted and therefore to a lesser extent.

Regardless of the type of nuclear explosion, the highly radioactive fallout particles, emitting alpha, beta and gamma radiation will fall to the ground at some point. In the case of surface and low altitude airbursts, nuclear fallout starts to

Figure 3: The mushroom cloud developing from a 21kt underwater nuclear test explosion named 'Baker', conducted by the USA at Bikini Atoll in 1946. The detonation was initiated 27 metres below the surface and was the first nuclear explosion to create fallout. The contamination was so bad, that of the >70 ships assembled for the test, all but 9 had to be sunk due to the extent of contamination, which turned out to be far worse than predicted. The islanders living on Bikini Atoll were evacuated before the nuclear test and were never allowed to return. To this day, the island is uninhabited, due to the radioactivity left behind.

rain down around 10 - 15 minutes after the nuclear explosion, although this will vary according to the size of the bomb, the height of detonation and the weather. Fallout will continue for around 24 hours, by which time, 50-80% of the fallout will have reached the ground. When it first arrives on the ground, the fallout will be extremely radioactive – due to the prevalence of short half-life isotopes. These will quickly decay, and so even 24 hours after the blast, radioactivity of fallout will have dropped significantly. The '7-10 Rule' or 'Rule of 7', established from empirical data, suggests that radioactivity of the fallout that reaches the ground will reduce by a factor of 10 for every 7 fold increase in the hours since the initial explosion – so 7 hours after the initial explosion, the radioactivity will reduce to 10% of the initial level of radioactivity, and 49 hours after the burst, it will reduce to 1% of the initial radioactivity. This is due to the exponential decay of the short half-life radionuclides and is the basis for the recommendation to remain indoors for two days before attempting to evacuate after being exposed to a nuclear attack.

Nuclear Winter – When a nuclear war breaks out between major powers, resulting in the detonation of many nuclear bombs, the resultant firestorms (such as those documented after the nuclear detonation at Hiroshima), are in theory capable of injecting huge amounts of soot into the upper atmosphere. With many such firestorms, the soot can have a very significant blocking effect on sunlight, and this in turn could reduce global temperatures by up to 20°C – in which case many areas would experience freezing conditions in the middle of summer, and agriculture would be reduced by up to 90%. This was highly popularised in the 1980s where a generation grew up under the threat of nuclear war. In the event of a nuclear winter, huge numbers of people would starve due to lack of food. In a severe nuclear winter event, mushrooms may be the only reliable food source, as they gain their energy by breaking down organic fibrous material, rather than using sunlight as an energy source.

It is important to stress that nuclear winters are entirely theoretical and have only been modelled with computers. Some simulations predict that the nuclear winter would last a few months, and others suggesting it would last for several years, but all serious efforts to model the climatic effect of a nuclear war have predicted a nuclear winter outcome of some kind.

Nuclear Tsunami – Some nuclear devices are intended to inflict destruction on a country by inducing tsunami waves. The 23kt 'Baker' underwater nuclear weapon test, carried out by the USA at Bikini Atoll, generated a column of water over 1 mile high, which created a 27-metre-high tsunami wave (as measured at the nearest monitoring station, 3.5miles away), caused by the water crashing back into the sea.

Tsunamis can affect any country with a coastline and very large tsunamis can travel several miles inland. In 2023 Russia started production of the Poseidon nuclear torpedo, claimed to carry a thermonuclear warhead up to 100Mt capacity (but more realistically, probably 2-5Mt). This device is designed specifically to create a large tsunami to destroy coastal regions, High profile figures in Russia have suggested that this device should be used to eliminate the UK. The device is claimed to produce a 1km high tsunami that could wipe out the entire UK or the entire eastern coast of the United States This is highly unlikely, as the tsunami would not travel far enough inland to achieve this. However, it would still cause a substantial amount of damage and could easily wipe out a major city such as London (with collateral damage to other nearby European coastal cities). Although the potential for such a strike would be of concern to residents of potential target cities, the fact is that the same size conventional nuclear weapon detonated in the air would have a far more destructive effect than one creating a tsunami.

Figure 4: Computer modelling has shown that millions of tons of soot injected into the upper atmosphere as a result of a nuclear war, would cause significant dimming of the sun, resulting is a big drop in global temperatures. Simulations have demonstrated that this could result in freezing temperatures in mid-summer for central belt USA, Europe and the UK.

When No One's Coming

4 NUCLEAR RISKS

Nuclear radiation is everywhere. The sky emits background radiation – originating from the sun as well as from distant stars and galaxies. Certain rocks such as granite produce radioactivity from their geological locations within the earth or, if houses are built with granite, from their walls or floors. When atoms of radioactive elements emit radiation, it is as a result of radioactive decay – and in some cases, this results in elemental transmutation – changing from one element to another. The element radon is produced from radioactive decay of uranium, which can be present in low concentrations in many types of rock. Such naturally produced radon presents a significant health hazard, because it is a gas, and can find its way through cracks and fissures into any underground space or indeed, into houses. This is particularly of concern in mines, where radon can accumulate to significantly higher levels. In fact, the biological effects of radon have been mostly identified and quantified from studies of miners. Being a gas, radon can easily be breathed in, and it has no smell, taste or colour, therefore it is very difficult to know that you are being exposed to it. It is thought to be the second most common cause of cancer (after smoking) and is considered the leading natural cause of cancer. Despite this, most people have not heard of radon, nor understand the risk factors involved.

In parts of the world, where the concentration of radioactive material in rock strata is higher than normal, natural radon levels pose a significant and real risk to health – seeping out of the ground into cellars and crawlspaces. For this reason, radon surveys are often carried out on sites of proposed properties or within existing properties, and if a significant risk exists, steps are taken to mitigate those risks – such as installing additional ventilation or extractor fans. In studies of some individuals, almost half of their annual radiation dose was caused by exposure to radon.

Radioactive materials are also present in certain household devices. For example, some smoke alarms (the 'ionisation' type, as opposed to 'optical' smoke alarms) contain Americium-241, an alpha particle emitter, though the amount is exceedingly small and perfectly safe in normal use. Older gas mantles – which are designed to provide a light source by burning bottled gas – contain the radioactive element thorium, and when used, can contaminate the air with daughter products such as radioactive isotopes of the elements radium, polonium and bismuth. Because of the perceived risk, modern gas mantles now use non-radioactive cerium instead.

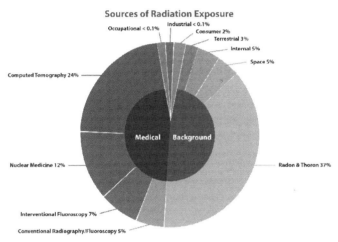

Figure 5: *Breakdown of Sources of Ionising Radiation affecting the American public. Note that although medical radiation constitutes quite a high proportion of the radiation exposure, this is due to high doses given to a small proportion of the general population, which is averaged over the population as a whole in the chart above.*

By kind permission of National Council on Radiation Protection and Measurements

Thorium is also used at low concentration in some TIG-welding electrodes, though again, alternatives are available based on cerium, although they are deemed not to have such good physical properties as the thoriated electrodes.

Other domestic sources of radioactivity include some ceramic glazes, which contain uranium, some older fluorescent light starters and radio valves, which contain weakly-radioactive krypton-85 or even radon gas, some 1970s and early 1980s desktop tape-dispensers, that contained monazite sand ballast – which is weakly radioactive due the presence of thorium and some camera lenses manufactured up to the late 1980s. Even dentures manufactured between the 1940s and the 1980s contained small amounts of uranium (up to 0.01%) in the porcelain, as the natural fluorescence helped make the dentures look whiter.

Probably the most dangerous source of radioactivity you are likely to encounter in a domestic environment is the radium-based paints used in some older watches or clocks. Radium, mixed with phosphors to create paint that glows in the dark, was regularly applied onto the hands and numbers of production watches from the early 1900s right up until the 1970s. These watches can actually often contain dangerously high levels of radioactivity. From the 1950s, tritium started to be used instead of radium, with lower levels of radiation, until these were eventually phased out in the late 1990s, being replaced by photo-luminescent paints, which did not contain radioactive material.

Although there have been concerning radioactive contamination incidents using everyday domestic items such as smoke alarms and watch faces, when we are talking about a nuclear incident, we are normally considering one of the following scenarios:

Nuclear Bomb

This is the most obvious source of radioactive hazard. During times of political tension, particularly involving superpowers, or involving nations with nuclear capability, this is especially worrying. Whereas nations and heads-of-state are generally aware that there are no winners in a nuclear war, there are situations where countries might initiate a nuclear attack regardless, for example if the country felt particularly threatened, or if the head-of-state has a point to prove or nothing to lose. Nuclear bomb detonations may be intentional or may be accidental – many stories exist of 'near misses' during the cold war, where one superpower believed, incorrectly, that the other had launched a nuclear attack. Due to the classified nature of such incidents, most were unknown until declassification at the end of the Cold War – and quite possibly other such events occurred, which were either not documented, or not declassified.

Thankfully, only a relatively small number of nations actually have nuclear bombs, but there are several less-stable nations who are trying to establish themselves as nuclear powers. Instability in such countries massively increases the risk of nuclear bombs being unleashed.

Dirty Bomb

For non-nuclear powers, or for small terrorist organisations, another option is the 'Dirty bomb' or Radiological Dispersion Device (RDD) – this is a bomb made up of conventional explosives mixed with radioactive material – potentially sourced from industry or medical science applications. A dirty bomb is designed to explode, dispersing radioactive material over as large an area as possible. The intention is to contaminate a location with radioactivity, rather than to harness the destructive explosive power of nuclear detonations. No nuclear detonation takes place in a dirty bomb. Fortunately, to date, no dirty bomb has ever been used as a weapon. In 1987 Iraq conducted dirty bomb testing, but the results were not deemed to be effective, so testing was discontinued. Whereas a dirty bomb has the capacity to produce significant contamination, requiring expensive decontamination procedures, it is not likely to cause many deaths, when compared to other terrorist weapons. It is, however, likely to cause widespread panic. Between 2010 and 2014, Israel carried out a series of experiments known as the Green Field Project, to determine how effective a dirty bomb would be, concluding that it was not of substantial concern and that the most significant effect would most likely be psychological.

Nuclear Plant Accident

There have been several serious nuclear plant accidents. The most famous being at Chernobyl in 1986, which contaminated around 150,000km^2 of land, 2600 km^2 of which will not be fit for humans to live in for the next 3000 years.

Japan's reactors at Fukushima also suffered a serious meltdown event in 2011, as well as exposing the highly radioactive material from the spent fuel storage ponds, as a result of a perfect storm of events initiated by a deadly tsunami, releasing radioactivity into the atmosphere, the sea, groundwater and contaminating land near Tokyo. The contamination was made far worse by the fact that the potential accident had not been properly planned for.

On American soil there have been a number of nuclear incidents; the most famous and significant being the Three Mile Island incident in 1979. In this case, the cooling system failed, followed by a failure in the emergency cooling system which resulted in the coolant draining, causing the fuel rods to become exposed, which subsequently overheated and caused a partial core meltdown. The operators had very little idea of what was going on, and this caused paralysis in their response to the unfolding crisis, subsequently leading to the release of significant amounts of radioactive gas into the air. No immediate deaths or injuries occurred, but the extent of radioactive contamination could have been very much worse, under different circumstances.

The UK has a track record of nuclear accidents – in 1957, a fire broke out in the core of the Windscale nuclear power plant and burned for 3 days. (Unfortunately, reactor cores are often made of carbon, which, when subject to high temperatures, can ignite and burn as a conventional chemical fuel) At one point, the blaze engulfed 11 tons of uranium. The 'Dounreay Shaft Incident' of 1977 in Scotland underlined just how inept the people in charge of nuclear reactors can sometimes be - in this case dumping liquid sodium and potassium into a deep, water-logged shaft whose purpose was (inappropriate) waste disposal. Most school-class chemistry pupils know that both of these metals react with water, which is exactly what happened, causing a build-up of hydrogen gas, which ignited and exploded, littering the surrounding area with radioactive material. This was, unfortunately, one of several nuclear incidents at the site.

The huge new reactor being built at Hinkley Point by EDF is of significant concern, as serious cracks caused by defective welding have been identified in the stainless steel containment vessels of several of EDF's other reactors of the same design, and a 2022 report has warned that Hinkley Point C could be affected by the same fault, potentially leading to nuclear accident. Sadly, these safety concerns are being voiced about a reactor even before it has actually yet been commissioned.

One aspect of nuclear plant accidents which has been largely ignored is the potential for targeting by terrorists. In the wake of the 9/11 terrorist attacks, people started asking whether nuclear power plants would withstand an impact from a commercial airliner. However, as this was never a consideration at the time most reactors were designed, it remains a significant potential risk, and several reports have identified this, although some contradictory reports claim that nuclear power stations would not be compromised in such a situation. Not only are the reactors themselves a potential target, but also the control rooms, spent fuel ponds, reprocessing plants, the nuclear fuel storage facilities and dry nuclear waste storage facilities. As well as terror attacks, nuclear plants may also become targets in war, as has been seen recently in the Russian invasion of Ukraine, in which there have been repeated incidents putting nuclear reactors at risk, including the containment building at the site of the world's worst nuclear disaster in Chernobyl and repeated shelling of the largest nuclear power plant in Europe at Zaporizhzhia during 2022 and 2023. Repeated disruption of its electricity supply cables, require emergency power generators to run, in order to provide an electricity supply necessary in order to maintain cooling pumps required to prevent a reactor meltdown.

Transportation Accident

Nuclear bombs, radioactive materials and devices containing radioactivity are all sometimes transported by road or rail. In the event of a road traffic accident, this may result in a potential nuclear emergency. However, extreme care is normally taken in transporting such goods, and containment in the event of an accident is normally built-in to the procedure, so the release of radioactivity is very unlikely to occur. Between the 11th and 14th of January 2023, a radioactive caesium-137 capsule smaller than the size of a small coin was dropped from industrial mining equipment being transported by road in Australia. Emergency crews searched over 800miles of road before it was finally recovered. The public was warned to stay away from the capsule, which would potentially cause radiation burns and radiation sickness at distances up to 5metres. It was recovered a couple of weeks later, on the 1st of February 2023.

Industrial/ Occupational Accidents

Some industries use highly radioactive sources. For example, in agriculture, radioactive sources are used for preserving fresh produce and improving their shelf life, mining companies use radioactive sources to measure the quality of the ores they are mining, likewise hospitals use them for various types of testing, imaging, and for radiotherapy to treat certain illnesses, such as to kill cancer cells, and highly radioactive sources are used for sterilising products from foods to medical equipment. Accidents involving such sources are generally not large incidents, but nevertheless happen regularly, and have resulted in deaths and contamination of many individuals and in some cases have resulted in contamination of neighbourhoods.

The most serious accident involving a medical radioactive source occurred in Goiânia, Brazil, where thieves looking for scrap metal stole a medical radiotherapy device from an abandoned hospital. As a result, several people lost their lives, several houses were demolished and topsoil was stripped from several sites, due to contamination. Over 100,000 people were screened, and 249 were identified to be contaminated as a result of the incident.

Figure 6: Abandoned gas masks left behind at a school in Pripyat – a ghost town of 35years, due to the evacuation caused by the explosion of Reactor Unit 4 at the Chernobyl nuclear power plant

When No One's Coming

5 HAZARDS CREATED BY A NUCLEAR EMERGENCY

In a nuclear explosion, hazards of various natures exist. Some are short-lived, and some may be very persistent, but to survive, you will need to avoid them all. The hazards are listed below, in roughly chronological order, although some will occur concurrently.

1) **Exposure to Intense light** – The initial flash of a nuclear explosion can, depending on distance, cause temporary blindness lasting around 40 minutes, due to bleaching of the retinas (up to 50+ miles from the detonation). In fact, nuclear explosions will produce a characteristic double flash – a very brief, very bright initial flash, followed by a second, gradually brightening flash. The first flash, lasting about a millisecond, is caused by the X-ray and ultraviolet emissions from the initial blast heating nearby air to around 1 million degrees Centigrade, which becomes ionised and briefly incandescent. Moments later, the fast-moving shock wave pushing outwards, causes the air in front of it to compress to

an extent that it, too, becomes very hot and ionised, and it is this light that produces the second, less bright flash. However, the hot, ionisation gases just ahead of the shock wave become opaque to light, preventing the light of the fireball within the boundary of the shock wave being seen. It is only when the shock wave begins to cool, that the light of the fireball starts to become apparent.

2)The **infra-red (heat) pulse** created in the initial flash can cause third degree burns (probably up to 5 miles from the detonation, depending on size of bomb, altitude of detonation and weather), and it will also cause many spontaneous fires involving any combustible material exposed to the heat pulse. The potential multitude of fires can cause secondary hazards such as thick smoke, low oxygen levels and high carbon monoxide levels – depending on your environment, this may make breathing difficult or impossible. As a result of the infra-red pulse, areas containing large amounts of combustible materials, can turn into firestorms, where strong winds are created by convection of fire gases, fanning the flames further. Firestorms are mainly an issue where there is a lot of combustible material in close proximity, such as forests, fields of crops, or towns built with wooden houses. A lot is written about nuclear firestorms, as it was a very significant effect after the nuclear bomb detonated at Hiroshima, for the aforementioned reason.

3)**Nuclear Electromagnetic Pulse** (NEMP)– There are actually three pulses caused by a nuclear explosion:

'E1' – this is the first, extremely brief pulse. It peaks approximately 5 nanoseconds (five thousand-millionths of a second) after the nuclear detonation, and decays away completely after 1 microsecond (1 millionth of a second). It is very high intensity and causes the most damage. It consists of a wide range of electromagnetic wavelengths from very low frequency radio waves right up to visible light. It is initiated by the intense gamma rays produced by the nuclear chain reaction. The gamma rays interact with atoms in the air, stripping electrons off these atoms in the air, and these electrons interact with both the earth's magnetic field and with the strong electric field caused by the nuclear explosion. These electrons crash into air molecules, resulting in the emission of electromagnetic radiation of various frequencies.

'E2' – this pulse immediately follows the initial E1 pulse, and is a secondary effect caused by scattered gamma rays and neutrons, starting around 1 microsecond after the detonation and ending around one second later. It is very similar to the EMP produced by natural lightning, and somewhat lower in intensity and therefore relatively easy to protect against (e.g., lightning arrestors and surge protectors).

'E3' – this is the slowest and longest lasting electromagnetic effect, with a duration of tens to hundreds of seconds. It is caused by the temporary distortion of the Earth's magnetic field resulting from interactions between the Earth's magnetic field and the E1 pulse. It can cause high voltages to be induced in long wires – such as power distribution cables, in a similar manner to geomagnetic storms occasionally produced by the sun. The E3 pulse can have a temporary effect on infrastructure but is not considered to be highly damaging.

Although the NEMP could theoretically have small adverse effects on the human body, you would need to be close to the centre of the blast to be affected (in which case you would be exposed to a highly lethal dose of radiation, and the blast itself). So, in practice, if you survive the initial explosion, it is highly unlikely that the NEMP will have any direct effect on you. However, the large electric fields caused by the nuclear explosion (up to a maximum of about 50,000V per metre - for high altitude detonations) can cause large electrical currents to be induced in metal objects or cables. These in turn can cause arcing discharges – like small scale lightning, and therefore there is a possibility of burns if you are struck by this arcing electrical discharge. This can be avoided by distancing yourself from metal objects if you are aware of an imminent nuclear strike. These electrical arcs and discharges are likely to only be significant close to the detonation site, where the gamma ray intensity and the shock wave constitute a far greater hazard.

The main concern with nuclear EMP is the potential for damaging electronics. Long wires, such as power distribution cables, will have very high voltage and currents induced, which can damage power distribution equipment and anything connected to the electrical distribution network, even at substantial distance from the detonation. It can induce high voltages in relatively short wires or electrically conductive metal – meaning that all electrical items are potentially at risk. If the electrical device is operational at the time of the EMP strike it is much more susceptible to damage, although damage can also occur in items that are not even turned on. Devices that are plugged into power outlets are particularly susceptible, even if the device or the power outlet is switched off. Semiconductors are also particularly vulnerable, as they are not particularly resistant to high voltages, unlike antiquated technology such as vacuum valves. The risk of EMP damage is much higher, the closer the equipment is to the point of detonation. Small handheld devices are more resistant and may not be destroyed by an EMP but will generally experience interference of some kind.

3) **Initial Blast** – the blast is the supersonic shock wave, which travels outwards from the centre of the nuclear explosion. This can cause buildings to collapse up to 3 miles away, and windows to shatter up to 10miles away.

To shield against the shock wave, you need physical protection – inside a strong structure such as a building is far better than being outside. It is safer still to shelter underground, where the blast would simply roll over the top of you - If you can take shelter in a basement, underground railway station or a bunker then you will escape the blast, although you are still at risk of being crushed by falling masonry from collapsing buildings. If you are sheltering inside a building, stay away from weak points such as windows, doors or the roof. If the building you are in is in danger of collapsing, take cover in doorways through supporting walls, where the structure is strongest. Cars provide little protection, but if you have no other choice, crouch down low in the car away from windows. If you are outside, seek any shelter you can, such as from walls, trees, or ideally natural dips in the ground or trenches, and lie face down with eyes closed until the radiation and blast has passed. If possible, use clothing as a crude mask to breathe through.

4) Exposure to radiation

Shielding and distance will reduce the amount of nuclear radiation that reaches your body.

Aim to put as much distance between you and the radioactive source as possible. The inverse square law applies to radiation and distance – so twice the distance reduces the exposure by a factor of 4. If you can increase the distance tenfold, you will reduce your exposure by 100 times.

Nuclear radiation is absorbed by materials. Thick layers of dense materials such as lead, steel, stone or earth provide the best shielding. Try to find locations with as much shielding as possible. The general rule is the thicker the shielding, the lower the direct radiation will be. Being underground is ideal – as you can be shielded in almost every direction. House walls provide good protection, especially if the walls are thick and made of stone or brick. However, there is very little shielding from above, as house roofs tend to be manufactured from timber beams with large gaps between them. The only effective shielding is afforded by the roof covering itself, which if made from slate, is usually two or three layers of a few millimetres thickness, or if made from clay or concrete tiles, 2 layers with an overall thickness of not more than 35mm.

5) Inhaling Radioactive Material

Fallout – Highly radioactive material made up of dust and radioactive gases are a serious contamination risk, and particularly hazardous through breathing.

Fallout will occur 10-30minutes after the blast. So even if you are outside during the explosion, you have time to get inside a building before the most harmful radioactive episode.

Fallout will gradually disperse and reduce over time. Most will have dissipated over the course of the first 24 hours. Even if you have good shielding, there will potentially be radioactive gases and dust in the air. Also, being a long way from the detonation does not necessarily mean you are safe – winds may push quite concentrated radioactive dust and clouds a long way. If you are downwind of the incident, you may still be at high risk at a considerable distance. Inhaling radioactive material is very dangerous, as radioactive materials may lodge in your lungs and airways and provide highly concentrated radiation to small areas of tissue for extended periods of time. The extent of airborne radioactivity is far more difficult to predict – due to wind, weather and topography. Some areas may be badly affected, whereas nearby areas may be spared. To minimise risk, you will need to seal off your building as best you can to exclude the outside air. One particular risk that can be mitigated to some extent is the effect of radioactive iodine-131 -which is frequently produced during a nuclear explosions or nuclear plant accidents (see Chapter 12 for further details)

6) Ingesting (Eating or drinking) Radioactive Material

Water supplies will more than likely be contaminated with radioactivity. Public broadcasts should inform you whether water is safe to consume. Any radioactivity getting into water supplies will be diluted to a low level, so even if you are advised not to drink the water, you can use it to wash yourself to decontaminate. If you do not know how safe the water is, do not drink it. Instead, drink bottled water or drinks from cartons. Similarly, do not eat any food that might have been exposed. Eat food that is sealed in containers, and wash/wipe off outside of the container before opening, to remove potential contamination from the outside. If you need to obtain food supplies during the incident, ensure that you only obtain food that is sealed inside packaging. Avoid fresh products that are not packaged or packaged in ventilated containers.

6 EARLY WARNING SYSTEMS & FALSE ALARMS

During the cold war, the USA and the UK operated a variety of early warning radar systems, together with other countries such as Canada, Denmark, Greenland and Iceland. The early warning system would provide the USA with 8 -15 minutes advance warning of incoming missiles, and the UK – being rather closer to the USSR - approximately 4 - 5 minutes advance warning – hence it became known as 'the four-minute warning' in the UK. The delivery system in the UK was based on the older air raid warning system of the Second World War, and consisted of local sirens, firework-style signalling rockets, church bells and even police whistles - to alert the general public of incoming nuclear missiles. Children growing up in the 1970s and 1980s lived in fear of the wailing sirens of the Four Minute Warning, which was often the subject of many unlikely playground myths. This warning system was run from 1953 until1992, when it was finally abandoned due to the collapse of the Soviet Union, signalling the end of the cold war.

In 1983, at the height of the cold war, German pop artist Nena, released the song '99 Red Balloons'. This song tells the tale of some children's balloons drifting into the sky, triggering an early warning system, followed by a chain of events leading to nuclear war. This very much captured the mood of the times, reflecting the fear and concern that existed, and went on to be a top hit all over the world, reaching the number one position in several countries.

Throughout the cold war, the USA operated and expanded a variety of ground-based radar and aerial monitoring systems (e.g. AWACS – Airborne Warning And Control System), until eventually satellite networks such as MIDAS (Missile Detection and Alarm System) took over, followed by a simpler and more reliable DSP (Defence Support Program) consisting of a network of geostationary satellites – extending the warning time from 8 minutes to something like 20 minutes. Today, the USA operates the UEWR (Upgraded Early Warning Radar), with radar stations based in the USA, Greenland, and the UK. This is operated by the US Space Force and provides continuous monitoring and early warning of impending attacks.

Delivering the warning

To be effective, early detection systems should provide a warning to the general public. In the USA, this would be done via television, radio, and cable broadcasts, as well as via the NOAA (National Oceanic and Atmospheric Administration) weather reporting radio system. In addition, members of the public can download the FEMA (Federal Emergency Management Agency) mobile phone app, which provides notification of local and national emergencies. Wireless Emergency Alerts (WEA) have been in use since 2012 and would also be used in the event of an anticipated nuclear strike. The WEA is a message sent directly to your mobile phone from the cellular mobile network, and arrives with a unique vibration and alert tone, to distinguish it from less high priority messages. Most mobile phones are capable of receiving these alerts. The UK government planned to introduce a similar mechanism in October 2022, although this was delayed to January 2023, and at the time of writing, it is expected to be tested nationally on the 23rd of April 2023, before going live. The system is known as 'Emergency Alerts', based on cell broadcasting, and will be automatically received by most phones based on iOS14.5 or later and by phones based on Android 11 or later. In fact, many phones based on older Android systems may also work. The UK system only works with 4G and 5G networks and will provide a 10second 'siren' alert sound to distinguish it from other types of messages.

False Alarms

During the Cold War, which existed between 1947 and 1991, the highest threat to worldwide safety was the prospect of a nuclear war between the USSR and the USA. Both nations were highly suspicious of each other, and over time this resulted in various escalations of tension between the two nations. Both countries used radar and early warning systems to notify themselves of an impending attack, and like most computer systems, these turned out to be fallible. At the time, public concern for an accidental nuclear war was high, and myths such as flocks of seagulls causing nuclear missiles to be launched were common. In reality there were probably more near misses than the public was aware of. The list of false alarms and near misses below, highlights how genuinely close the world had come to nuclear war, and demonstrates how easily this could happen accidentally, particularly through reliance on computers and automated processes.

1956

American NORAD air defence system detected a number of events which was initially thought to be signs of a Russian Nuclear Attack. These were: a British bomber downed (which turned out to be as a result of technical issues), Soviet MiG fighters over Syria (which were escorting the Turkish President home from Russia), an unidentified plane over Turkey (which was in fact a flock of swans), and unusual activities by the Russian Black Sea Fleet (actually, a scheduled training exercise).

1960

Greenland's air defence radar mistakenly interpreted a moonrise over Norway as a large scale Russian nuclear attack. Although the supposed attack was taken as a serious threat, it was noted that the Russian leader was in New York at the time, and it was deemed unlikely that the Russians would strike the USA whilst their president was visiting the country.

1961

An American bomber aircraft broke up in the air above North Carolina, dropping two armed 3-4 Megaton nuclear bombs in the process. Only a simple low-tech switch prevented a huge catastrophe – something that could easily have ended differently if, as a result of the accident, there had been a short circuit in one of the components.

Strategic Air Command Headquarters simultaneously lost contact with NORAD, and multiple ballistic missile early warning sites. This was interpreted as a coordinated attack, and all available air forces were readied for immediate take-off, until aircraft already in the air confirmed that it was a false alarm.

1962
During the Cuban Missile Crisis, when relationships between the USA and the USSR were highly strained, an alarm was sounded when an intruder (actually a bear) was detected climbing a security fence at Duluth Sector Direction Center (an USA air defence centre at the time), an alarm was sounded, and due to a fault at another base in Wisconsin, Air Defence Command (ADC) ordered nuclear-armed F106 interceptors into the air. The pilots had been informed that there would be no practice alert drills, and so they fully believed that a nuclear war had been initiated at the time. There was a very real possibility of these interceptors shooting down ADC bombers which were in the air at the time, whose presence was classified and secret at the time. The interceptors were called off at the last moment by an officer driving onto the runway as they were about to take off.

A couple of days later, a Russian submarine narrowly avoided launching a nuclear-armed torpedo at American warships, when it had been surrounded by the USA navy during manoeuvres near Cuba. The Russian captain lost contact with Moscow for a few days and believed that the Americans were dropping depth charges as an aggressive attack on the submarine. (In fact, the American ship was dropping signalling depth discharges, intended to indicate to the submarine that it should surface and be identified) Instead, the submarine captain believed that war had broken out, and ordered the nuclear torpedo to be launched. However, the chief of staff refused to grant permission, and persuaded the captain to surface instead and make contact with Moscow, averting potential disaster.

Also at this time, an American U2 aircraft was shot down over Cuba, and another strayed into Russian airspace due to a navigation error, prompting the USSR to scramble jets to intercept it. The USA also scrambled interceptors to escort the U2 back into friendly airspace, each of these interceptors being armed with small nuclear warheads. The pilots of these planes had full control of the launch of these nuclear missiles, and the situation could easily have escalated into an actual nuclear confrontation.

1965
The command centre of the Office of Emergency Planning experienced a power cut which affected a large chunk of North-eastern America. This caused the triggering of several detectors which interpreted the power failure as being the result of a nuclear attack.

1967
A powerful solar flare hitting Earth's atmosphere adversely affected the NORAD air defence system, giving the impression that it had been jammed by the USSR as part of a nuclear attack strategy. As a result, nuclear bombers were nearly launched for a counter strike.

1969
North Korea shot down an American military plane flying 90 miles from the North Korean coast. The US President Nixon, allegedly drunk at the time, gave the order to prepare to retaliate with nuclear strikes against North Korea, but the mission was stood down a few hours later before it had taken off.

1973
During the Yom Kippur War, Israel found itself in a precarious situation where Arab forces might overrun Israel. As a result, they requested an Airlift from the USA, which did not immediately happen. Therefore, Israel ordered the arming of jets with 13 nuclear missiles, to stand by, ready to launch a nuclear attack. When the USA discovered this, they initiated an airlift immediately, thus preventing a nuclear attack unfolding.

1979
Computer errors in various divisions of American air defence organisations, led to a full-scale initiation in response to a suspected incoming attack from the USSR. The NORAD computer had reported an incoming attack of 250 missiles heading for the USA, requiring confirmation of a counterattack within about 5minutes. In the meantime, NORAD computers upgraded the threat to 2200 missiles, and nuclear bombers were prepared for take-off. Approximately 7 minutes after the warning, it was confirmed as a false alarm, after verification by satellite and radar systems. The false alarm had resulted from a training scenario being inadvertently loaded into the computer. Several similar false alarms were raised over the following few months due to electronic faults.

1980
A Soviet submarine launched four missiles as part of a training exercise. One of these was deemed to have a trajectory aimed at the USA. An emergency meeting was arranged at which point, it was deemed to be a non-existent threat.

1983
Shortly after a Korean airliner had been shot down over Soviet airspace, an alarm was raised by a Soviet computer, indicating five incoming nuclear bombs from the USA. A Russian Lt. Colonel refused to order retaliatory missile strikes, and convinced his superiors that it must be a false alarm, his reasoning being that the Americans would not launch an attack of just five missiles. It turned out he was correct, and the false alarm was caused by the new Soviet early warning system being confused by sunlight reflecting off high clouds. It does raise the question, however, of whether he would have suspected it was a false alarm if the computer had erroneously reported several hundred missiles rather than just five.

Also in 1983, a joint British/American exercise intended to simulate the response to a nuclear attack, fooled the Soviet Union into believing that NATO was preparing for a large-scale nuclear attack. The Soviets responded by preparing nuclear-armed bombers and nuclear ICBMs (Inter Continental Ballistic Missiles) in readiness for retaliation.

1991
During the Gulf War, Iraq launched Scud missiles at Israel and Saudi Arabia. At the time, there were serious concerns that Saddam Hussain would use Weapons of Mass

Disruption (WMD), such as chemical weapons, and he was warned by Israel, the USA and the UK, that any attack using such weapons would invoke a nuclear response from Israel. As a result of these warnings, and because Iraq did not feel sufficiently threatened, WMDs were not launched against Israel, but if it had chosen to do so, there is a strong chance that a nuclear attack against Iraq would have been initiated in response.

1995
Russian submarines were put on high alert to ready for a retaliatory strike, when radars detected what was thought to be an incoming nuclear warhead. In fact, this was a Norwegian rocket for studying the Northern Lights phenomenon. Russia had been forewarned about the research rocket, but the information had not filtered down to the radar operators.

As you can see, history is littered with false alarms and near-misses throughout the period of the cold war. There may well be other examples which have remained secret and will perhaps never become public knowledge. It is all too obvious that the combination of owning a nuclear arsenal, together with the fear generated by Cold War suspicion and secrecy on both sides, may lead to accidental launches of nuclear strikes. Whereas it was fairly clear that neither side was likely to initiate a nuclear war, both sides would not hesitate to use nuclear weapons in retaliation for a nuclear strike initiated by the other side, which left the door wide open to accidental nuclear Armageddon, where neither side had any intention of starting a nuclear war but started a war accidentally by retaliating against an imaginary threat.

During the Cold War, this mutual suspicion and distrust was regarded as the most likely potential cause of an all-out nuclear war.

Figure 7: In Europe and the USA, early warning of incoming nuclear missiles would be broadcast on TV, Radio and Internet. Until the 1980s, they also used a system of sirens in the USA, and in the UK the old Second World War air raid sirens system would have been used right up until the early 1990s, when the system was finally dismantled. Most nuclear power stations still use sirens as a warning system to notify the public in case of a significant nuclear accident.

When No One's Coming

7 PERSONAL QUALITIES REQUIRED FOR SURVIVAL

During an emergency, you are dealt a hand of cards based on pure luck. Sometimes your luck will be poor, sometimes it will be good. However, the process of maximising your chances of survival, given the hand of cards you have been dealt, is the product of your knowledge and training and of your personal qualities. These qualities are essential for staying calm and focused, in the face of adversity, and for staying resilient, given the potential gravity of the situation. The list below indicates the qualities which are beneficial or useful to ensure your survival.

Resilience

Resilience is a highly important quality – the ability to bounce back from adversity, to remain strong in the face of tough challenges and the ability to recover from setbacks. In the aftermath of a nuclear war resilience will be essential.

Survivors will face the physical and emotional toll of the disaster and will need to find new ways to proceed through the desolation that surrounds them. Those who have the ability to cope with the stress and trauma surrounding them, will be better able to survive in the long term.

Adaptability

Being able to adapt to changing situations is another key quality required for survival. In addition to the dangers and hazards presented in the aftermath of nuclear war there will be chaos and unpredictability, and survivors will need to be able to adapt quickly to changing circumstances. They will need to secure safe food and water supplies, they may need to locate or create shelters, and may be faced with unfamiliar terrain. Those who are able to adapt efficiently to their new surroundings, will be more likely to navigate an unfamiliar world, more likely to survive, and to rebuild their lives in the aftermath of the disaster.

Physical Fitness

Life in the aftermath will be tough. You will need to work hard to survive, both mentally and physically. You may need to carry heavy items, sometimes over long distances. You may need to remove or reposition house debris and fallen masonry, or you may need to dig your way in or out of somewhere. In the event of violent confrontation, you may need to run to escape. If you are physically fit, you will be best placed to achieve all of this.

Courage

Courage is the ability to face fear and take action in spite of it. During a nuclear war, there will be many dangers and uncertainties. Those who are courageous are better able to face these challenges head-on and take the necessary steps to protect themselves and their loved ones.

Determination

Some people give up easily. Some will try for a while and then give up. Some people possess the single-minded determination necessary to complete a task or challenge, regardless of how difficult it is or how long the odds, and regardless of how many setbacks or wrong steps they have taken. Those with this level of determination will be more likely to complete the many necessary steps required to safeguard survival.

Self-reliance

In the worst-case nuclear scenarios, when you really need help, you can be sure that no one's coming. You will therefore need to step up and take care of yourself and your family, independent of outside help. To compound problems, in a nuclear war, there may be disruptions to communications and transport networks, and supplies of food and water may be inaccessible, intermittent, or non-existent. Those who are self-reliant, and not dependent on outside help will be much better equipped to provide for themselves and for their family throughout the disaster.

Survival Instinct

Those possessing a 'will to survive' will find a way around seemingly insurmountable obstacles and survive when others fail.

The qualities described above, give an indication of the type of attitude you will need. In addition to these qualities, survivors of a nuclear war need to be resourceful, patient, and able to work with whatever tools and resources they have, with others or, if necessary, alone. Resourcefulness will provide creative solutions to challenges arising out of nuclear war, and patience will help deal with the slow, painstaking effort required to make progress towards rebuilding normal life - or something that bears some resemblance to normal life. Although the ability to work with others may seem to be an odd requirement, a charismatic person will be more likely to diffuse aggression on the part of other survivors and may help draw people together to work for the greater good, and to achieve more than might be possible from the sum of the individual efforts.

Surviving a nuclear war requires a combination of personal qualities in addition to knowledge and training, with huge advantages for those who have carried out some kind of preparation. Of these qualities, the will to survive, resilience and adaptability are probably the most important. By cultivating such qualities, and preparing yourself for the worst, in a sensible, measured fashion, whilst remaining upbeat and positive, you will achieve the best chance of surviving potentially catastrophic events that the future may bring.

8 WHAT TO DO IN A NUCLEAR EMERGENCY: I) BEFORE THE EVENT

Some people will want to prepare for their safety during a nuclear emergency 'just in case'. Some will prepare if prompted, some will try to prepare when a nuclear emergency is imminent, and some will not prepare at all.

When to prepare

If you are considering making preparations for your safety, there are some warning signs which may prompt you that this is a sensible course of action. For example, an increase in world unrest, particularly in your own country or neighbouring countries, or between world superpowers that hold nuclear arsenals may be an indicator – particularly if you live in territory controlled by one of those superpowers.

Location

There is a joke about a tourist asking a local for directions to a nearby town. The answer given was 'Well I wouldn't be starting from here'! This has some relevance to choosing the location of your home. If you live near a military base, centre of government, a busy port, the coast, a major city, important infrastructure or a power station, then your risk may be considerably higher than if you were to live in a sparsely populated area away from such locations. You could make impressive preparations to cope with disaster, but if where you live puts you in the firing line, then it may be counterproductive. Admittedly, most people would not easily consider changing the location of their home to reduce their risk, as it is often strongly tied to factors such as workplace, schools, other family members, social groups, or because they have lived in that location for a long time, but if you have the luxury and inclination to choose a home anywhere, then you can plan to minimise your risks. Both the UK and the USA have well-defined nuclear targets – military bases, factories supplying military hardware and major cities, government buildings and infrastructure. Recently, this has even been plotted as a map by a major house-seller to help house-buyers decide which part of the country provides the safest option.

In *Figure 8*, I have used a Russian Cold War list of UK targets, and plotted the likely contamination zones and fallout plumes for each of these targets. Looking at the map, it is obvious that little of the UK landmass is immune to the after-effects of a nuclear confrontation. Likewise, in the USA, vast tracts of land would be rendered uninhabitable in the event of a nuclear war with Russia, as the map in *Figure 9*

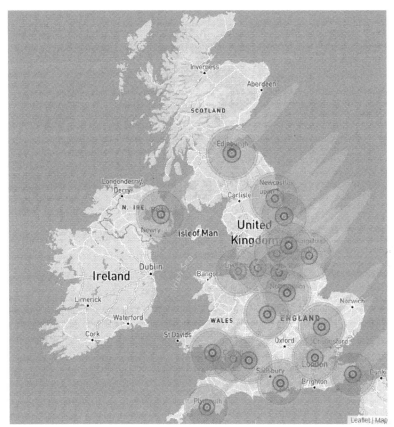

Figure 8: Potential Russian Nuclear Attack on the UK.
I have plotted this map to show c.40mile contamination zones plus fallout plumes for hypothetical nuclear detonations at Russia's "Top 20" UK city targets, according to a declassified 1970s document from the cold war era, which was drawn up by the UK Government and circulated amongst defence chiefs. In addition to the 20 cities featured in this map, the document lists 23 RAF bases, 14 USAF bases, 10 radar stations and 13 Royal Navy bases, totally 106 targets in total. Only detonations on the towns have been marked on this map i.e., 20 out of 106 prospective targets - Obviously, if all of these additional targets were included in the map, there would be very little of the UK unaffected by nuclear contamination, in the event of a nuclear war with Russia.
Map generated using NUKEMAP by Alex Wellerstein (https://nuclearsecrecy.com/nukemap/), Map data © OpenStreetMap contributors, CC-BY-SA, Imagery © Mapbox

shows. This map was produced by FEMA (Federal Emergency Management Agency).

Even if you feel that a nuclear war would not actually occur, if you live close to a nuclear power station, then you may be concerned about the potential for nuclear accidents, such as the ones that occurred in Fukushima, Chernobyl and Three Mile Island. During the Chernobyl disaster, all people were evacuated within a 30km radius. However, the area of land contaminated by radionuclides was far larger than this. The cost of evacuating and potentially relocating people was very much a consideration, and the USSR was heavily criticised for the time it took to initiate the evacuation. The Three Mile Island incident had a relatively small evacuation zone – only 2½ mile radius, however plans were afoot to evacuate up to 20miles radius depending on how the emergency panned out, and in fact, many residents evacuated from towns up to 40miles away as a precaution.

At Fukushima, an exclusion zone of 20km was established, and residents up to 4km were evacuated. Again, a far larger area was affected, with contamination detected as far away as Hawaii (1500km). The US Nuclear Regulatory Commission recommended that residents up to 50miles from the Fukushima Reactor evacuate.

The map of the UK shown in *Figure 10* has been plotted to show 40mile radius contamination zones around the nuclear power plants across the country – large areas are potentially under threat should a serious nuclear incident at a power station occur.

Of course, nuclear contamination very much depends on the

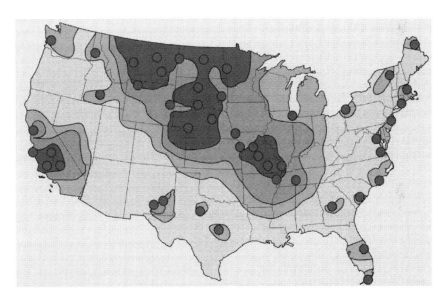

Figure 9: Map of the United States of America, produced by FEMA (Federal Emergency Management Agency), providing a colour-coded image showing areas of low contamination (lighter colouring) to high contamination (darker colouring), following hypothetical 1980s Russian strikes of primary Cold War targets. The darkest colours are considered areas subject to lethal amounts of fallout, whereas the yellow areas are deemed to be relatively fallout-free.

wind strength and direction – contamination will be much worse downwind, and stronger winds will push the contamination plume further from the source. It is impossible to predict how bad an accident will be until it actually happens.

The UK is particularly vulnerable, being a relatively small country, so approximately 98% of the population lives in an area where there is some potential threat of nuclear incident.

Preparation of Supplies

There is no right or wrong in terms of how much preparation you should carry out. However, in the event that a disaster occurs, people will always wish they had prepared more, regardless of how much effort they have made. There are simple things you can do, which are beneficial in other ways – for example, everyone should have a First Aid kit to hand, regardless of whether they believe preparing for a nuclear disaster is worthwhile. By spending a little extra on buying things not normally included in First Aid kits, and by attending a basic emergency training class, you will expand your capacity to deal with a medical emergency regardless of its cause. Likewise, you may decide that it is worthwhile to stockpile extra food and drinks. There are many situations where this can be useful – flooding, heavy snow, or natural disasters such as forest fires, landslides or earthquakes. It may also be useful even if there is not a disaster – for example, if you were incapacitated due to illness or an unexpected lack of transport.

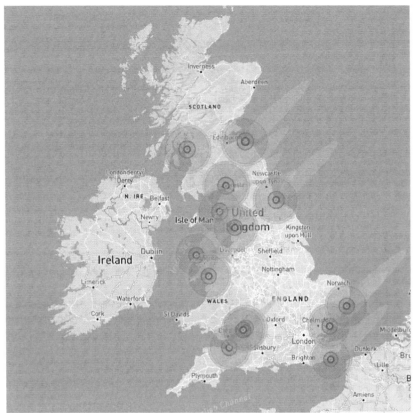

Figure 10: Potential Nuclear Contamination after a major incident at any of the operational nuclear powerplants in the UK.

This map plots c.40mile contamination zones plus fallout plumes. which might occur if any of the UK's nuclear powerplants experienced a catastrophic event such as those that occurred in Chernobyl or Fukushima. There are other locations that have not been represented on this map, such as fuel reprocessing plants, research or military reactors, and reactors undergoing or awaiting decommissioning.

Map generated using NUKEMAP by Alex Wellerstein (https://nuclearsecrecy.com/nukemap/), Map data © OpenStreetMap contributors, CC-BY-SA, Imagery © Mapbox

Shelter

If you have the time, space and inclination, you may decide that the construction of a nuclear shelter is worthwhile. This does not necessarily need to cost a lot of money, but it will involve a fair amount of work. A basement in your house could provide an ideal location, or you could build a simple shelter outside. Large investments of time and/or money in something that is hopefully never going to be needed are difficult to justify, but if you can use it for another purpose – such as a 'man cave', recreation area, or additional bedding area for guests, then it may seem a worthwhile investment after all.

Make a Plan

The most important thing is to make a plan – not necessarily on paper, but at the very least, in your head. A template is provided in Appendix I to help you prepare your own plan. To do this, you will first need to decide what you want to achieve, and how far you wish to go. You may wish to add a page for each aspect of your plan, where you can go into much greater detail. Decide how much you can afford to spend in terms of finance, either an absolute amount, or a weekly or monthly amount that you can use to develop your plan over time - and stick to it. Remember, Rome was not built in a day, and nor does your plan need to be. Start off with something rudimentary, and build on it, as time and finances allow.

Within your plan, set out your aims and who you wish to protect – obviously your immediate family members will be on this list, However, you will probably want to add in your pets, and maybe extended family members if they live close enough to be reachable during an emergency. You may wish to help your neighbours, and also people that you do not know at all. However, you also need to be pragmatic – prioritise your list – people who must be protected by the plan, and those you will try to protect if you have the resources to do so. You need to determine which scenario(s) you wish to protect them from (some will automatically be covered if you opt to prepare for a more extensive disaster – for example, if you prepare for nuclear war, by default, you will also be covered if you are cut off by heavy snow for a week. Some people will not consider a nuclear war to be a serious threat, whereas due to their location, the prospect of being cut off by heavy snow for weeks at a time may seem almost a certainty. Once you have decided on the scenarios you wish to entertain, you need to plan what you will do, how you will achieve it, and for how long you will be able to do so. If you live in an area that is occasionally subject to flooding, it would be entirely sensible to prepare for being cut off for up to a week due to floodwater. However, if your plan is intended to protect you during nuclear war, it would hardly make sense to prepare for a disaster lasting just one week.

Remember your plan needs to be able to adapt to changing circumstances, and you will need to build in flexibility to do so. If there is a local emergency, your safest option might be to evacuate to a relative's or friend's house temporarily, and of course, you can return the favour by offering your house to accommodate them, if the situation is reversed. Of course, if staying in or near your own home is a safe option, then you are able to create a much greater useable emergency resource, than if you have to leave your home.

Determine your Assembly Points / Shelter Locations

You need to plan for different scenarios – if your building is destroyed, then you need an outdoor assembly point. If your building is intact, and you need to get inside urgently, your assembly point should be a central ground floor room furthest away from outside walls if possible, or ideally the basement. If you have a bunker at home, then that is an obvious location, unless access is blocked.

Establish your Evacuation Plan

If, on the other hand, you need to evacuate, then you need to plan for your potential destinations – it might be a friend or relative's home at a location safely away from the disaster, or it might be a local community centre, or you may decide to rely on the authorities to provide you with shelter.

For each person in your plan, prepare a suitable escape bag – this should be a sturdy rucksack with, as a suggestion, the following items: maps, compass, high energy snacks, drinking water, identification documents, change of clothing, waterproof coat, protective gloves, blanket, eating utensils, shatterproof plate and mug, toothbrush and toothpaste, a few weeks supply of personal medications, water purification tablets, face mask and filters (FFP3 grade),torch, 2-way radio and broadcast radio receiver, foil blanket, basic first aid kit, sharp scissors, money, mobile phone charging cables and portable charging device (solar panel or hand-crank device), wet wipes, female hygiene supplies, a couple of toilet rolls, a small towel, hand crank broadcast radio receiver a whistle, knife with 4-6 inch blade, foldable shovel etc. You should also ensure that each person has some form of time-keeping such as a watch or mobile phone. If a person does not routinely carry such a device, then one should be included in the bag. For children, think about what comforts they might need. Whereas a teddy bear might seem a waste of valuable space, if it helps a young child cope with an extreme situation, it is a valuable addition. If children are worried about leaving their possessions, explain that you intend to return for them later, after the danger has passed.

Plan your Evacuation Route

If you must leave your home, how do you intend to do so? By foot is the slowest, and most arduous option as it involves the most physical effort, and you will be limited in what you can carry.

If, on the other hand, you plan to take your car, as well as your escape bags, you will be able to load it with a good quantity of food, clothing, and equipment. But if the roads are blocked due to an obstruction or the sheer weight of traffic, you must be prepared to abandon your car and most of your supplies, taking just your escape bags. Of course, if you have a four-wheel drive vehicle, you may have the option of driving off-road to escape the traffic jam. Whilst in a vehicle, beware of carjackers and thieves who may try to take your vehicle or the supplies inside it. Drive with windows wound up and doors locked.

Motorcycles, bicycles, electric bicycles ('e-bikes') and electric scooters may have a big advantage in that they are able to filter through large traffic jams, are able to take routes not available to wider traffic. Most can travel on-road and off-road with ease, can pass through restrictions that would prevent access to most motor vehicles, and can even be carried or hoisted over, around or under obstacles. Electric bikes have a big advantage in that they can be quicker, can cope more easily with uphill inclines, and can have a decent range - typically 40 - 100miles, although at higher speeds the range will be reduced. Even when the battery is exhausted, you can continue with pedal power alone. If one or more of your proposed evacuation routes is over rough ground, and you plan to use a vehicle of some kind, make sure it has suitable tyres, suspension and ground clearance to cope with the route that you intend to take – and if possible, test it. Remember that access along off-road routes may be very much worse after significant rainfall, so take this into account.

Study local maps or aerial views, and actually travel the proposed routes – whether they are paths, tracks, open moor or roads. Check to see if these routes become grid-locked during rush hour at busy times of day – if they are, then you can bet they will be grid-locked when the population is panicking and trying to leave in large numbers.

Consider when you will evacuate. Road traffic will be highest shortly after a serious incident. In a nuclear emergency, you are usually best sheltering at home for the first 48 hours anyway. When you have taken the decision to evacuate, think about how many other people will be making the same choice. It may be better to evacuate during the night or very early in the morning – using cover of darkness to aid discretion, and hopefully being a time when the traffic congestion will have eased.

Establish your Communication Plan

If you need to evacuate, ensure that your family remain together and travel as a group. There are advantages to this – in terms of security and being able to help one another if problems arise. If you need to traverse difficult terrain or obstacles, a group of people may do so more easily than a person alone, as they are able to help one another overcome the obstacles. However, it is important that you have a plan for remaining in touch in case you get separated – either deliberately or accidentally. The easiest way to achieve this, is with two-way radio, using mobile (cellular) phones as a fall-back (as the cellular network may not be functional during a disaster, or it may be overloaded due to sheer numbers of calls).

Make sure every family member has a radio and a mobile phone. Make sure the devices have charged batteries, spare batteries, and if you are planning for a nuclear bomb emergency, make sure they are adequately protected against EMP (see Chapter 13). Decide on a channel (or frequency), and make sure everyone knows which channel to use. Have a backup channel planned, so if there are other people using the channel you intended to use, everyone knows which channel to change to, but should only do so if the original channel is busy. It may be that the chosen channel will only be busy for a few minutes, in which case, you can simply wait a couple of minutes and try again. Plan how you should formulate your messages – for example, you may decide to attempt contact for a period of 10minutes, by alternating between calling and listening on the radio channel. You may decide to speak in code to prevent identification by unknown third parties (or, in fact, you may benefit from asking for help of anyone listening, as they may have picked up a radio signal from a lost member of your family, that you cannot receive at your location). Have a plan for when to try to make contact – for example you may opt to try to make contact immediately if you get separated and continue to do so for 2 hours. After that, you may try to make contact on the hour (e.g., at 1pm, at 2pm, at 3pm etc) and continue to do so for three days. After 3 days, you may decide to restrict contact attempts to twice a day for example at noon and at 7pm for two weeks. By doing this, you maximise your chances of making contact early on, and continue to keep the options open as long as possible, without detracting too much from your time which may be required for other matters.

Make sure everyone knows the principles of radio communication: that line-of sight between the two antennae provide the best chance of making contact over long distances, and that radio waves – particularly those of higher frequencies such as the VHF and UHF bands – are easily absorbed by metal, concrete and buildings – so the range you achieve in the middle of a city will be far less than the range you will achieve in flat, open areas. To achieve maximum line-of-sight to the unknown location of a missing person, be aware that elevation is highly advantageous. Even a pair of small transceivers of 0.5W or less may make contact over 20-50miles, if one or both of them occupies an elevated position, which puts far fewer radio-frequency absorbers, such as buildings, cars, tress or topography in the way of the line of sight.

Establish your 'Shelter in Place' plan.

The safest option may be to shelter where you are, especially if there has been a recent detonation, and fallout has occurred or is likely to occur. You need to plan how and where you will do this – in a basement, a room central to the house, or in a bespoke bunker or shelter – either built and equipped in advance or built during the emergency. You need to plan how you will secure the building if it has been damaged, and how to seal it to reduce the inflow of contaminated air. If you plan to build a shelter during an emergency, make sure you know how to do so, and what materials you plan to use, and where they will come from. Plan the dimensions of the shelter so that it is suitable for the number of people who will use it, plus sufficient storage space for the supplies they will need, and make sure you have the space available to build it.

Make out a list of the essential tools and consumables you need to achieve your aims and ensure these are stored in the place you plan to use as a shelter, or in a bundle somewhere nearby, where you can quickly grab them at a moment's notice. (See Chapter 14 for suggestions of useful tools).

Checklists

Make a checklist of what you need to do, both in advance of, and under the circumstances of scenarios that you are planning for.

This should be your plan - not mine, and not one that

someone has written for you. Modify and shape the plan to suit your situation and to plan for scenarios that you feel are important. - If things are important to you – include them. If they're not, then do not feel you should include them because someone else says you should. Any plan is better than no plan, and any preparation is better than no preparation.

Finally, make sure you have considered a time frame for reviewing and revising your plan. Your circumstances and those of your family may change – your location, your jobs, new family members, or family members growing older or leaving the household. You may find you need to include new medication in the plan, or take account of a new mobility issue, or you may have new options that were not available or not affordable previously. Items that are in storage need to be checked on a regular basis, so failures or expiries can be noted, and exchanged. You should check everything to make sure it is there; it is not damaged and has not failed – maybe once or twice a year. This also helps familiarise yourself with the equipment and supplies you have and helps prevent you from forgetting what you have or how to use it. It would be very easy to make very thorough preparations, and then by not revisiting it for a long period of time, be unable to take advantage of life-saving aspects of your plan because you have forgotten essential information.

8 WHAT TO DO IN A NUCLEAR EMERGENCY: II) IMMEDIATE ACTION

The earlier you react, the safer you will be. So, if you know that a nuclear threat is imminent or has just occurred, you should take action immediately.

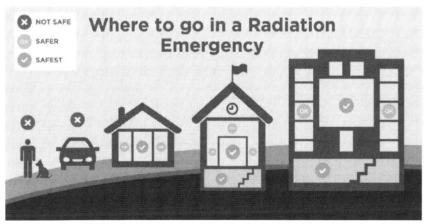

Figure 11: FEMA poster very clearly indicating the best and worst locations during a nuclear attack.

1) GET INSIDE

Make your way to the nearest solid building. Whilst outside cover your nose/mouth with material to try to reduce the amount of radioactive dust you are breathing. If you can use a damp cloth, then this will be more effective than a dry cloth. The sooner you can get inside the better – ideally within a few minutes. You need as much concrete/brick/steel between you and the radiation as possible. If possible, get into a basement – this is the best-shielded part of a building. If you are in a large building move to the centre and stay as far from the outer walls as possible, and avoid being near windows, doorways or roof - where shielding is worst. If you can take cover underneath a table or behind a sofa, this will help a little – as many injuries are sustained by splintered glass or wood when the shock wave hits a building. When you enter the building, immediately remove any outer clothing, taking care not to disturb or distribute any contamination. Carefully bag it to contain the contamination and place the bags well away from people or pets. Wash any exposed skin/hair thoroughly with soap and water to remove contamination on the skin and put on clean clothing if available. If you do not have access to water, use a wet wipe to wipe down any exposed skin and hair. Do not use conditioner when washing hair, which will potentially increase your exposure to radiation, as it can help fallout particles stick to the skin or enter pores in the skin. Make sure pets are inside and keep them inside. Decontaminate pets by washing with soap/water or using wet wipes. Close and lock all doors and windows.

2) STAY INSIDE

Do not leave the building. You may be concerned about friends and relatives, but do not attempt to locate or join them. Instead, they should make their way to the nearest solid building and remain there. You should not try to meet them nor encourage them to travel to you. Avoid being outside, as that will exposure you to the highest levels of radiation. Cars offer very little protection, so don't assume that being in a car is keeping you safe. Stay inside a building for at least 24hours, longer if possible – to give the radiation chance to disperse.

3) STAY TUNED
Listen for Radio, television, internet broadcasts giving instruction and further information. It is impossible to predict how a radiological emergency would play out, as it depends on many things – wind, rain, type of incident, local topography etc. In the event of an emergency, it will not be obvious how serious it is or when it will be safe to venture out. Therefore, once safely inside, listen for public service broadcasts which may offer further instruction. Keep as many lines of communication open as possible – internet / mobile phones / TV / Radio. Depending on how badly damaged the public infrastructure is, it is likely that some forms of communication will not be available. Radio is likely to be the most reliable, but some or even most channels may not be available. Scan the whole radio band if you cannot locate a broadcast station. Try again later if necessary. If you have a shortwave radio, listen with that too. Also, bear in mind that history has taught us that public bodies tend to downplay the risks of nuclear accidents – so make your own judgement as well.

When No One's Coming

8 WHAT TO DO IN A NUCLEAR EMERGENCY: III)SHORT TERM ACTION

Assess your level of exposure and the danger you are in. If necessary, initiate first aid for any individuals that are showing signs of medical emergency (see First Aid section)

Once you are safely inside, you may need to stabilise your environment to provide ongoing protection.

Firstly, seal your building to reduce the rate that contaminated air enters. Use tape/plastic/polythene to seal off vents such as bathroom extractors, kitchen extractors, heating vents, air bricks near fireplaces, close air controls on solid fuel / wood burning stoves, seal off the flue on open fireplaces, trickle vents in window frames or poorly fitting windows and doors. Any broken windows can be sealed off with tape/cardboard/cling film/timber or whatever material is available. Do this as soon as possible.

Water supplies may initially be clean enough to drink but are likely to end up contaminated. If water is available from the building's taps, fill your bath with water to provide a reservoir in case water infrastructure fails, as it will still be useful for washing and decontamination. Fill as many buckets and containers as you can, sealing off the top with a lid or cling film in case you need to use it for drinking water at some point. However, unless you have been informed that the water is safe to drink, use bottled water, fizzy drinks or fruit juices in the short-term. Ration the water – do not waste it, as you do not know when you will be able to access it again. Avoid the temptation to use rainwater, which is likely to be highly contaminated, or any water which has been stored in an exposed container outside. Also, be aware that if you boil water contaminated with radioactivity it will NOT make it any less harmful.

Initiate daily iodine supplements – See table in appropriate section of this manual for dosage details. Continue daily supplements for as long as radioactive iodine risk exists.

Electrical power is also likely to be unreliable or non-existent. Make sure you have torches, candles, a lighter or matches. If the weather is cold, you will need to think ahead and arrange plenty of blankets and warm clothing in advance of night-time. Be aware that your heating system is likely to rely on electricity, so when the power goes off, your heating will stop working too. Make sure you locate your batteries, and if they are rechargeable, make sure they're all fully charged. Even if you have plenty, ration your lighting at night – as you do not know how long you will be without power.

8 WHAT TO DO IN AN EMERGENCY: IV) MEDIUM TERM ACTIONS

If you have not done so already, pack a case of your essential belongings, but only take what you can carry. Take important papers – passport, driving licence, credit cards, cash. Take essential equipment – face masks, radiation monitors, broadcast-receiving radio, mobile phone. Ensure children have your details with them, in case they get separated. Bear in mind that depending on the level of contamination, you may not be allowed to take anything at all.

If you have time, hide any valuables that you are leaving - to protect it from looters. If the incident is small, you may be able to return and recover these later.

When advised by public authorities that it is safe to do so, evacuate to a safe area, well away from the contaminated area. Stay away for as long as possible, and do not return until the authorities say it is safe to do so.

Take particular care to source ongoing food/drink that is not contaminated throughout the incident. Many people who survived the nuclear bombs dropped at Hiroshima and Nagasaki drank rainwater to quench their thirst, and unknowingly subjected themselves to considerable additional nuclear contamination. Depending on the severity, it may be necessary to source uncontaminated food and drink, if at all possible, for years. (See Chapters 15, 16 & 17).

STRATEGIES FOR SURVIVAL
9 HOW TO DECONTAMINATE

If you have become contaminated with radioactive material, or if you suspect so – for example because you have been outside during or after a nuclear explosion, you should follow the decontamination procedure described below.

If you are helping to decontaminate someone else, you should do so using adequate protection to avoid becoming contaminated yourself – i.e., you should wear a face mask to prevent inhalation of radioactive dust, goggles to protect your eyes, and disposable gloves to protect your hands. You should also cover all exposed skin to avoid radiation burns, and wear a disposable coverall with hood, if one is available.

1. Remove outer layer of clothing.

Most of the radioactive contamination will be on the outer surface of your outer clothing. So carefully removing this will instantly reduce your level of contamination by 80-90%. If you have disposable face masks, wear one during this part of the procedure to minimise the risk of breathing in radioactive particles. Remove clothing as soon as possible, but do so slowly and carefully, to avoid disturbing any dust. You may not be able to see radioactive dust, but even very small quantities can be extremely toxic, so just because you cannot see it, does not mean it is not there.

Remove any valuables or personal possessions from pockets, together with any jewellery or piercings and seal in a small bag such as a zip lock bag. If multiple individuals are being decontaminated, you should label each bag to identify the owner.

If at all possible, cut clothing away from your body with scissors rather than undressing - this can reduce the chances of disturbing radioactive particles on the surface of your clothing and reduces the extent of inhaling harmful material. Take particular care not to pull contaminated clothing over the head, as this would increase the chances of inhaling radioactive particles. Place the contaminated clothing in a large plastic bag. Avoid filling the bag with air, and when sealing the bag, do not squeeze the air out of the bag – this will simply blow radioactive material out of the bag and contaminate the surrounding area, and create an ingestion hazard. Seal the bag with a cable tie/tape. Alternatively, a sealable container such as a drum or a plastic container with tight-fitting lid may be used. Place the bag in a location away from people and pets.

2. Radiation Survey

If you have a radiation survey meter, use the meter to do a whole-body survey, slowly passing the meter over every part of the body at a distance of around 1cm. If time permits, continue by surveying any radioactive shrapnel wounds, followed by any open wounds, then facial openings – nose, mouth, ears and eyes. If any areas of contamination are discovered, mark around these with permanent marker for easy identification during decontamination.

3. Wash off

Ideally wash off under a shower, using plenty of soap and water. Note that the soap will likely become contaminated during the process, so, assuming it is liquid soap, decant what you need into a secondary container and use it from that, disposing of the secondary container at the end of decontamination. If you are using a bar of soap, cut off a section of the bar for use in decontamination, so that the remainder can be retained for the future. Use warm water, as cold water will cause skin pores to close up, making it more difficult to clear contamination from within the pores, hot water will cause the blood vessels in the skin to dilate, and will increase the rate of absorption of contamination through the skin. Any mild soap is suitable - you do not need to use special 'radiation soap' as may be claimed elsewhere. If possible, when using a shower, direct the water away from the body, rather than allowing the contaminated water to run down the body. Wash thoroughly, taking care not to scrub, scald or cut / scratch skin, and rinse your eyes and ears – where radioactive dust may settle. Blow your nose to remove any radioactive dust that may have been filtered out by your nostrils and rinse out your mouth with clean water. Wash your hair with soap or shampoo. Do **not** use conditioner, as it can help radioactive materials adhere and avoid being washed off.

If you are unable to take a shower: Assuming running water is available from taps (even if it is considered unfit to drink due to contamination), wash your face, hands, hair and any exposed part of the body. Use soap and plenty of water. If no running water is available, then use moist disposable wipes (or damp paper towel) to wipe clean any exposed parts of the body. Where possible, wipe in one direction, towards extremities rather than back-and-fore. Ensure that hands and face are cleaned well. Wipe eyelids, eyelashes, ears and around and under fingernails. Seal the used wipes in a plastic bag, as for contaminated clothing.

4. Repeat

Perform another two cycles of decontamination immediately afterwards, using clean wipes/towels each time, to gradually reduce the contamination level. Conduct a whole-body radiation survey at the end of each cycle, aiming to reduce contamination level ideally to less than twice the background level. If this is not possible, reduce to the lowest practical level possible. It is not normally recommended that you carry out more than three decontamination cycles.

Keep any cuts and abrasions covered during washing to avoid getting radioactive particles into the wound and wash separately afterwards. If it proves impossible to decontaminate some areas (e.g., if the radioactivity has migrated into the skin layer), do not scrub vigorously as it removes the outer dead layer of skin which may be protecting the living skin underneath. Instead cover the contaminated areas in a dressing, to prevent it from being spread, ingested or contaminating others. Replace dressing every two days, unless the wound is infected, in which case the dressing should be changed daily If the dressing becomes soaked with blood or pus or any other liquid, it should be changed immediately.

5. Open wounds/ Radioactive Shrapnel

Assume all wounds or shrapnel have heavy contamination unless the survey meter says otherwise. Wounds will provide much greater potential for internal contamination, so great care should be taken. If contamination level is high, mask off surrounding skin with waterproof material to protect from radioactive run-off and use the shower head to rinse off and irrigate the wound as much as possible. Use long forceps (or alternative grasping tool if that's all that is available) to gently tease out small items of shrapnel. Instruments used will likely end up heavily contaminated, so treat them as such – bag them for safe disposal and remove from habitation areas. Substantially irrigate the wound after removal of shrapnel. This will need to be done several times to remove absolutely all internal contamination. Remove any visible fragments of radioactive material from wound if present, using forceps. Do not remove deeply penetrating items from wounds - this will likely cause profuse bleeding, potentially leading to fatality. Therefore, take all possible steps to seek medical help if faced with deeply penetrating injuries.

In cases of high-level contamination, it may be necessary to remove heavily contaminated flesh from the wound, the scope of which is very much outside of this book.

6. Dress in clean clothes

Clothes that have been stored inside a drawer or cupboard, away from radioactive contamination are safe to wear. If you do not have safe clothes to wear, use a blanket, towel or even bin bags with holes cut for arms and head. Due to the likelihood of issues with heating systems, it is as well to select warm clothes and plenty of layers, unless the weather is particularly mild.

As a last resort, you can re-use your contaminated clothing if the contamination is not too severe, by shaking/brushing off any contamination outside, ideally rinsing them in some water to reduce the contamination. Ensure full face mask, gloves and disposable suit is worn during this procedure. However, this option should only be considered where no other options are available, and it must be accepted that a significant and ongoing contamination will occur.

10 MONITORING RADIATION LEVELS

There are various methods of monitoring radiation levels. Most people are familiar with the Geiger counter, which is based on a low-pressure tube system and still the commonest method in use today. However, there are now semi-conductor alternatives. Radiation monitors normally rely on the ionising properties of the radiation as a mechanism of measurement.

Figure 12: How a Geiger Counter works.

Radiation monitors are also known as 'survey meters' – these are intended for measuring ambient radiation levels, or the radiation being emitted from a radiation source.

Dosimeters also measure radiation, and are generally fitted to outer clothing, like badges, and measure the actual dose or total amount of radiation that a person is exposed to – these are cumulative, in other words they add together the total amount of radiation they are exposed to – so the reading increases as time goes by. By subtracting the start reading from the end-reading, you can establish how much radiation a wearer has been exposed to. Some radiation monitors will also double as a survey meter – meaning they can be used for both purposes.

Measuring radiation

Radiation is measured in a variety of confusingly named units – Becquerel, Rem, Roentgen, Curies, Sieverts, rad, Gy – some are units of radiation dose, others are units of biological risk from exposure.

We should focus on the unit 'Sievert' ('Sv'), which is an internationally agreed standard of equivalent absorbed dose of radiation – in other words, it provides an indication of the level of seriousness of a dose of radiation, regardless of which type of radiation a subject has been exposed to For a sense of scale, a chest x-ray would be equivalent to receiving a dose of approximately 0.1mSV (the 'm' stands for 'milli', meaning one thousandth of a SV). A human's typical annual background radiation dose from natural sources might be in the region of 2-3 mSv/year, but this increases up to an average of around 6mSv/year in the USA, when other sources of radiation are taken into account such as internal exposure (from low level radioactivity in food and drink), medical exposure (from x-rays and CT scans), air travel (as there is less protection from solar sources of radiation at higher altitudes), exposure from industrial process, exposure in workplace and from consumer goods . The dose also varies according to where a person lives, as a result of natural variation in background emissions from geological sources and artificial sources such as the processing or utilisation of nuclear fuels. Radiation monitors tend to display hourly radiation rates – so a typical background radiation level under normal conditions would be 0.05 - 0.1µSv/hr. In a nuclear war, however, the worldwide background level will increase substantially, even at places remote from the detonations. A dose of 4 to 5 Sv received over a short duration would have a 50% chance of killing an average human within a month. There is no 'safe' limit for radiation, but the lower the radiation levels, the less dangerous it is.

How to use a Radiation Monitor

A radiation monitor reading Sv/hour would give an indication of how safe/dangerous an environment is. In practice, you would need to be near a dangerously high level of radiation to measure on this scale, so most monitors will give readings in µSv/hr (micro-Sieverts per hour – 'micro' meaning one millionth). Background radiation levels are typically around 0.05 – 0.2 µSv per hour (3.5mSv/year). When reading radiation levels be very aware of the units – micro (µSv), milli (mSv) and Sv.

1,000,000 μSv = 1,000 mSv = 1 Sv

When using the radiation monitor, turn it on and let the reading stabilise. Then you can move the radiation monitor to areas that you wish to test, letting the reading stabilise before taking the reading. When investigating sources of radiation, or suspected sources, you should hold the monitor a few inches away from the source. If the source is particularly powerful (e.g., if you are searching for highly radioactive debris on a floor, it is a good idea to fix the monitor to a pole of some sort, so that you can keep more parts of your body at a safe distance. This is particularly true because of the inverse square law that applies to radiation – basically if you double the distance, you quarter the amount of radiation. So, at a distance 10x further away, you will receive one hundredth of the radiation dose.

If you are surveying an individual for potential contamination, make sure you are wearing adequate protective equipment, and pass the survey meter over all parts of their body, at a distance of about 1cm, and without actually touching the individual or their clothing. Move the survey meter slowly, to give it sufficient time to pick up the radioactivity, so that you do not miss anything.

Many radiation monitors emit a 'click' each time an ionisation event is sensed in the detector head. This is a very useful feature, as it is very easy to hear whether the radiation levels are increasing or decreasing, without needing to look at the meter the whole time. Under normal background radiation

Figure 13: Radiation Survey Meter in Operation.

levels, the instrument will click once every few seconds, but at high levels of radiation, the interval between the clicks will reduce, and may even merge into a continuous crackling sound. More expensive machines will have selector switches to decrease the sensitivity of the clicking so that it can still be a useful trend indicator in high radiation environments.

11 FIRST AID

It is highly recommended that you take a first aid course. If this is not possible, read some dedicated first aid publications, and watch online videos, but practical training is far, far more effective. First aid is a highly useful skill that everyone should have, and decent training will make you confident to deal with situations that you would not think you could handle previously. Buy a book dedicated to first aid, read it, and store it with your emergency supplies.

Also, you should prepare a decent first aid kit. Suggestions for contents are as follows:
- Selection of plasters – large and small
- Conforming bandage
- Large triangle bandages
- Safety pins
- Micropore tape
- Sterile Wound dressings (various sizes)
- Sharp Scissors
- Fingertip pulse oximeter (displays pulse rate and

blood oxygenation level, by clipping onto fingertip)
- Antiseptic cream
- Burn gel
- Tough-cut medical scissors (for cutting off clothing)
- 36 inch (96cm) flexible splint
- Surgical gloves
- Disposable face masks
- Foil blankets
- Small roll of clingfilm (for burns)
- Hand sanitiser / Sterile wipes
- Haemostatic dressing (for controlling severe bleeds)
- Pressure dressing (for applying pressure where you cannot use a tourniquet)
- Bag Valve Mask ('BVM' – normally used with oxygen, but can also be used without, if oxygen cylinder not available)
- Tourniquet
- Set of Oropharyngeal airways (OPA) / Nasopharyngeal airway (NPA) plus lubricant
- Glucose gel
- Chest seal (for puncture wounds to chest)
- Valved Chest Seal (as above, but able to vent air to prevent lung collapse)
- Suture kit (with sterile needle/thread)
- Constipation medication / Diarrhoea medication
- Pain killers
- Antihistamine (for controlling allergies)

Below you will find brief summaries of significant first aid procedures which you might need to perform in an emergency, but it must be highlighted again, that is not substitute for proper training.

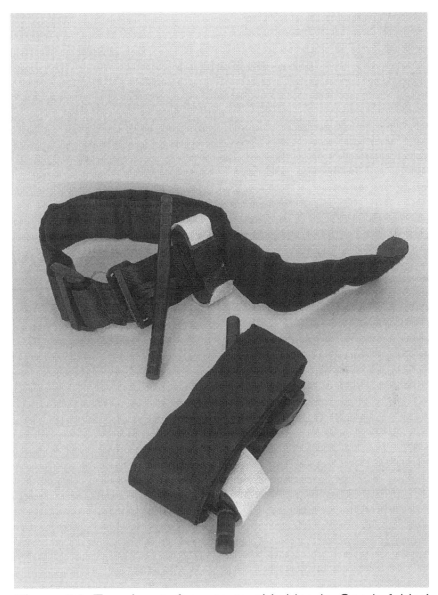

Figure 14: Tourniquets for catastrophic bleeds. One is folded for storage, and the other is opened and ready for use. These are readily available online and are not expensive. Purchase in pairs, as you often need to apply a second tourniquet to completely stem the blood flow from a wound.

Initial Assessment

If the casualty has serious or life-threatening injury, contact emergency services immediately, if at all possible. You may need to carry out lifesaving first aid, such as tourniquet a catastrophic bleed, or perform CPR on casualties that are not breathing.

Unconscious Casualty

Before you start, consider your safety and that of the casualty. Is it safe for you to approach the casualty? If it is not safe, is it possible to approach the casualty only for as long as it takes to drag the casualty to safety? Or do you need to deal with another issue before you can approach the casualty - for example if the casualty location is on fire, you may not be able to approach the casualty at all, without first dealing with the fire. If you have to drag the casualty to a location where it is safe to assess and work on them, then this must be the primary consideration. Do not try to lift or carry the casualty. Drag them by whatever you can – clothing, or under their arms, or by their wrists, or their feet. If you cannot do this you can use a loop of rope – place both casualty's arms through the rope loop, then draw one side of the loop through the other, behind the casualty's head, so that you now have a loop to drag them with, which is secured under each arm, going behind the head and not restricting the casualty's neck. In an emergency, you can make a very useful tool to drag a casualty, by cutting a seatbelt out of a car, and knotting it together to form a loop. This will spread the load, so as not to bite into your hands or into the casualty's body, during extrication.

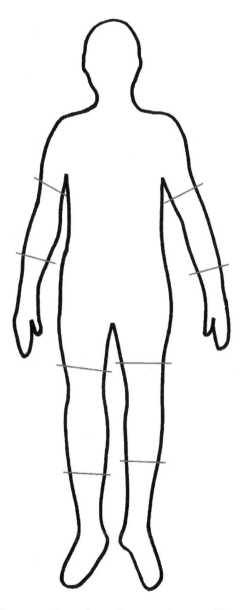

Figure 15: Common locations for tourniquets. Ensure tourniquet is at least 2" (5cm) away from wound in the direction of the torso. If necessary, a second tourniquet can be applied above the first. Write the time of application on the tourniquet and on the patient's forehead or cheek.

The (c)ABCDE of Resuscitation

When you and the casualty are in a safe place, you will need to go through the **cABCDE** process – Catastrophic bleed / Airway / Breathing / Circulation / Disability / Exposure:

Catastrophic Bleed

Check for catastrophic bleed. Catastrophic bleeds are typically associated with deep wounds, crushing injuries, gunshot wounds or accidents relating to vehicles and machinery. However, not all catastrophic bleeds are immediately obvious, so you will need to check. A person can bleed out within a few minutes, so if they are gushing or pumping blood out, this needs to be dealt with urgently.

Firstly, you should expose the wound and apply sustained, direct pressure - as hard as you can - to the wound to stop the bleeding, unless this is not possible due to an impaled object. (Do not remove deeply embedded objects). For severe wounds, if you have haemostatic dressing, pack the wound with this material as tightly and as quickly as possible, before applying firm pressure. Haemostatic dressings contain a blood-clotting product that works independently of the body's own blood-clotting process and are extremely effective in controlling severe bleeds. If the blood is pumping out, then you will need to use a tourniquet immediately. If a haemostatic dressing is not available, use a dressing or any cloth material that is to hand. Do not release the pressure for 15 minutes – to give the blood time to clot. If the blood starts to soak through the material, apply another on top – do not remove the first dressing. If at all possible, elevate the area relative to the rest of the casualty as soon as possible. Do not lift the cloth off the wound – which may cause the forming blood clots to fail. If bleeding stops, tie the makeshift dressing on to the wound by wrapping around with tape or strips of fabric – whatever you have available. If this does not work, you can clamp the limb with your hands at key pressure points to restrict blood flow. This basically squeezes the artery against bone (which will be painful to the casualty). The main pressure points for the legs are located in the centre of the front crease of the join between the leg and the groin. You will need to drive the heel of your hand or your knee into this area, to slow down bleeding from a lower limb. For bleeding from an upper limb, you will need to clamp the artery on the inside of the upper arm, located under the biceps muscle. Use your fingers or thumb in the grove between the biceps and triceps (muscles at the front and back of the humerus – the upper arm bone) to achieve this. Other pressure points for controlling bleeding manually (NB these are NOT positions for applying tourniquets) include the inside of the elbow, to control bleeding in the lower arm; at the back of the knees for bleeds in the lower leg and at the wrists and ankles for controlling bleeds in hands or feet. Remember that raising the bleeding limb so that it is higher than the heart will allow gravity to assist you If the blood is pumping out under pressure, then you will need to use a tourniquet.

If someone else is able to help, you can apply pressure to a pressure point to stem the flow of blood whilst they search for something suitable to use as a tourniquet. If you have a dedicated first aid tourniquet, then use this. If not, use any strong material that can be tightened without tearing, such as shirts, garment sleeves, towels, a sock or a good bandage. Some belts are suitable as makeshift tourniquets, but stiff belts will not be suitable, as they will not twist easily, Shirt ties, shoelaces, power cords and zip/cable ties are too thin – tourniquets should be at least 4cm in width. Any narrower, and there is a good chance that it will cause nerve and tissue damage as well as unnecessary additional pain. However, if nothing else is available, and direct pressure to the wound is not working, you will need to use anything that can be tied off tightly. Apply the tourniquet at least 2-3inches (5-7cm) above the catastrophic bleed i.e., between the wound and the torso, and if possible, apply it directly on top of skin, rather than over the top of clothing Tourniquets can be applied to upper or lower limb, but must not be applied over a joint (knee/elbow/wrist/ankle), as it will be ineffective in these places. Whatever material you are using, tie it in a loose-fitting loop with a secure knot and pass something through the loop to act as a windlass. Something like a wooden spoon, cutlery, a strong stick, or a pair of scissors is ideal. Then tighten the tourniquet by winding the spoon or a strong stick, so that the tourniquet starts to twist and tighten. Continue to do so until the bleeding stops (some dark residual blood oozing slowly from the wound is acceptable). If the casualty is conscious, this will be extremely painful, but you must continue the process to save his or her life. If the tourniquet is not 100% effective; you should apply another tourniquet a few inches higher up. Use something to secure the windlass to stop it unwinding – such as wrapping and securing another bandage or strip of clothing material over the top of it. Once you have successfully applied the tourniquet, write the time of application on the tourniquet itself, and on the patient's forehead or cheek, in the form of a T, followed by the time it was applied (e.g., 'T -13.25')– this is essential information for Emergency Service Personnel. Do

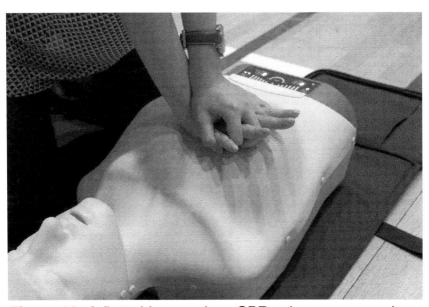

Figure 16: A first-aider practices CPR using a mannequin.

not be tempted to remove or loosen the tourniquet. This should only be done by medical professionals. Tourniquets can remain place for two hours before tissue damage starts to occur.

Airway
Moving on, you will need to check if the casualty's airway is clear. If you need to, roll the casualty over, so that they are face-up, and tilt their head back slightly, so that the airway is not restricted. Open their mouth and check that their airway is clear and there is nothing blocking it. If necessary, use a finger to remove anything that is in their mouth. Do not feel any further than you can see. If the casualty has vomited, roll them to their side to remove the contents of their mouth, to avoid choking.

Breathing
Check for breathing. Lower your head so that it is slightly above theirs and look down towards the chest, you should feel or hear their breathing and you should see the chest rise and fall slightly. If not, the casualty is not breathing.

Circulation
Check for a pulse. If someone is injured, their pulse may be very weak, and it may be difficult to locate in their wrist. The easiest way to check for a pulse is by pressing a couple of fingers at either side of the neck, under the jaw to feel for a pulse in the main carotid arteries.

If the casualty is not breathing and has no pulse, then you need to start CPR <u>immediately</u> (described below). CPR stands for Cardio-pulmonary Resuscitation. (Heart – Lung resuscitation). If someone has a cardiac arrest, (i.e., their heart stops beating,) then CPR is urgently necessary to keep them alive. For every minute that it is delayed, their chance of recovery reduces by 10%. Rescue breaths ('mouth-to mouth') are not necessary with CPR – the chest compressions will cause air to move in and out of the lungs to some extent.

Disability
Check to see if the casualty is responsive. Try calling them loudly – is there any verbal response? If not, check to see if they react to pain – push thumb firmly into their forehead, or pinch the muscle halfway between the shoulder and neck, or perform a sternal rub (grinding motion with hand on patient's sternum (bony section in middle of chest). If the casualty does not react at all, they are deemed unresponsive. Try to establish the reason for this – are they diabetic and require sugar (sugar gel can be rubbed into the gums of an unconscious diabetic casualty, to help restore their blood glucose levels). Have they had a bad response to a food or drug? Or experienced a fall or an injury? Or been electrocuted? Have they been overcome with smoke? Or fainted? Or had a head injury?

Exposure
In order to assess the casualty properly, and to reveal any hidden injuries, it may be necessary to fully expose them. To do so in an emergency situation, it is easier and simpler to cut through their clothes rather than trying to remove them intact. It also avoids the potential to cause additional damage, as you may not know how or why the casualty is incapacitated. Respect the casualty's dignity and ensure that the casualty is covered back up as soon as possible to prevent excess heat loss. If you have a blanket, use it to

cover the casualty, and if the casualty is outdoors and it is cold, get something underneath the casualty to insulate them from the cold ground.

When the casualty is exposed, you can look for any injury, signs of bleeding, trauma, puncture wounds, needle marks, bleeds etc. As well as looking, feel for any injury (and a pain response from the casualty). You are feeling for fractured bones –the tissue around a break may be swollen, or the bone may feel like it is in the wrong place. Pressure on even a cracked bone will normally provoke a pain response from a conscious casualty.

CPR
There are two versions of CPR: 'Hands only' and 'CPR with Rescue Breaths'.

Hands Only CPR
Immediately call the emergency services (if available) or shout to an onlooker or passer-by and ask them to do so whilst you get started with CPR – every second counts. If you do not have help, after dialling the Emergency Services number, set the phone to loudspeaker mode, so that you can continue the phone call, hands-free – that way you can start CPR whilst talking to the Emergency Operator. Tell them you are carrying out CPR, and do not worry if you cannot answer their questions immediately. They will wait. They can also guide you, if you are unsure of what to do.

Make sure the casualty is lying on his/her back, facing upwards, and make sure they are lying on a firm surface. If they are on a bed or sofa, or other soft material, quickly drag them onto the floor so that the process is effective.

Kneel alongside the casualty and place the heel of one of your hands on the casualty's breastbone (centre of the chest), and place your other hand, palm-downwards, on top of the first hand, interlinking fingers.

Lean forwards so that your shoulders are directly above your hands, and start compressions, using your full bodyweight to bear down. You should see the casualty's chest rise and fall by 5 – 7cm with each compression stroke. After each stroke, release the pressure completely, to allow the casualty's chest to return to its original position. You should aim to provide around two compressions per second. It will feel fast, and it will be hard work. It will also feel somewhat brutal – you need to push hard to provide sufficient force to massage the heart itself. There is a fair chance that you will break ribs, although obviously this is not the goal of CPR. About 70% of CPR casualties suffer broken ribs as a result of CPR, and most receive multiple fractures. It is, however, an acceptable price to pay for being resuscitated. As you tire, you will unconsciously start to reduce the force. For this reason, if anyone else is available to help, take turns, swapping over every two minutes. Continue with the compressions until help arrives, or until the patient regains a pulse. If you know the location of a public defibrillator, dispatch someone immediately to obtain it, and when it arrives, turn it on and follow the instructions (the instructions are usually given audibly and are also often written on the machine itself). You may need to provide CPR for a long period of time, so do not give up after a few minutes. The main objective of chest compressions is to keep the casualty alive until you can get medical help. Sometimes casualties will start breathing again of their own accord – if this happens, then stop CPR immediately, and start running through the cABCDE checks again.

It should be borne in mind that if a casualty requires CPR, their situation is extremely worrying. Unfortunately, only a small percentage of casualties requiring CPR survive – around 10%, so do not be surprised if you are not successful, but by giving CPR you are providing the best chance of survival. Ideally you should keep going until emergency medical help arrives. Many people will ask when they should give up. This is a difficult question to answer, and different medically trained staff will give different answers. You should certainly continue for at least 20minutes (or until the patient returns to spontaneous non-assisted breathing). Some studies have concluded that you should give CPR for at least 40minutes. There are several reported incidents where casualties have made a full recovery from CPR lasting hours, including a couple that endured CPR for 5 – 6 hours.

CPR with Rescue Breaths
This process is very similar to the previous version of CPR, but also provides supplemented breathing:

Make sure the casualty is lying on his/her back, facing upwards, and make sure they are lying on a firm surface. If they are on a bed or sofa, or other soft material, quickly drag them onto the floor so that the process is effective.

Kneel alongside the casualty and place the heel of one of your hands on the casualty's breastbone (centre of the chest), and place your other hand, palm-downwards, on top of the first hand, interlinking fingers.

Lean forwards so that your shoulders are directly above your hands, and start compressions, using your full bodyweight to bear down. You should see the casualty's chest rise and fall by 5 – 7cm with each compression stroke. You should aim to provide around 2 compressions per second. After 30 compressions, stop and give the casualty 2 rescue breaths. To do this, tilt the casualty's head back, and lift their chin. Their mouth should naturally fall open. Pinch the casualty's nose and seal your mouth over theirs. Blow steadily into their lungs until inflated (take care not to over-inflate, particularly in the case of children) Then repeat the process – 30 compressions, 2 rescue breaths. Keep repeating until help arrives or until the casualty starts to breathe for themselves.

Radiation Sickness

Symptoms of radiation sickness will exist if casualties have been exposed to very significant levels of radiation and can include confusion, disorientation, nausea, diarrhoea and vomiting, or bleeding from gums.

Vomiting may be an indicator of exposure to very high levels of radiation. The shorter the interval between exposure and vomiting, the higher the exposure experienced by the casualty. The onset of symptoms can be anything from minutes to days, depending on severity. The casualty may recover quite quickly from the initial symptoms and be apparently healthy for a number of days (known as the 'walking ghost' phase), after which the initial symptoms will set in again, potentially also with seizures and coma. The casualty may become very ill, and will be prone to infection, which is due to the damage caused by radiation to the immune system. This phase can last anything from a few hours to several months, followed by a slow recovery, or potentially, death.

Treatment:
1)As the casualty to remove clothing and shoes. Do not pull clothing over head or step out of clothing as this will increase exposure. Instead cut through clothing to remove it. Removing clothing will remove 80-90% of external contaminants. If the casualty is unable to do this without help, then person carrying out this procedure is liable to become contaminated, so, if possible, have the casualty remove their own clothing. If you need to assist, wear a facemask, gloves and protective clothing. Bag up all removed clothing and protective clothing to reduce the opportunity for re-contamination.

2)Wash casualty with copious quantities of soap and water. You do not need to use any special radiation-soap, ordinary soap is very effective. Find the casualty uncontaminated clothes or a large sheet/towel to wear. Likewise, decontaminate yourself, as you are quite likely to have been contaminated during this procedure.

3)If the casualty has potentially been exposed to Iodine-131 contamination, start iodine supplements immediately. Be aware that the casualty will be more prone to infection over the coming days, and care should be taken to clean any wounds to minimise risk. The casualty may be suffering from headaches or diarrhoea, in which case over-the-counter medication to treat these symptoms may be effective. Dehydration is a significant risk, and the casualty should be encouraged to drink sufficient fluids. If the casualty is in pain, provide painkillers if available.

Burns

Thermal burns ('heat burns') and radiation burns are both likely injuries following a nuclear detonation. These have similar symptoms, except that heat burns are apparent immediately after exposure to excessive heat, whereas radiation burns can show up anything from 3 hours after exposure to several days after exposure. People near Chernobyl who were unaware of the nuclear disaster, mistook their radiation burns for sunburn initially, as the burns were experienced by people who had been outside in the sun, and had very similar symptoms. Severe radiation burns can include blistering and ulcers. Skin may begin to heal, followed by a return to the reddening and blisters. Complete healing can take several weeks to a few years, depending on radiation dose. Infection is a major concern after radiation burns, due to the damage potentially extending well below the skin, together with the adverse effects on the immune system caused by radiation of bone marrow – where immune cells are created before entering the bloodstream. Severe radiation exposure can also lead to temporary hair loss, which may take several weeks to recover from. Casualties who experience radiation sickness or having significant radiation burns should ideally receive prompt professional medical care. They also need to avoid any potential sources of infection, as well as getting plenty of sleep and eating a healthy diet. For severe cases, casualties may need blood transfusions or bone marrow transplants.

Burns used to be classified into 1st, 2nd, 3rd and 4th degree burns. Now, the terms 'superficial', 'partial thickness' and 'full thickness' are used to describe the extent of burn injuries.

Superficial (1st degree): Only the outer layer of skin is affected. Skin is red, painful but not blistered.

Partial thickness (2nd degree): Lower layers of skin are affected. Skin is red, very painful, blistered and often swollen.

Full thickness (3rd degree): The entire thickness of the skin is affected, there is little or no pain as nerve endings are destroyed, and the burn may appear white or blackened.

4th Degree burns: These penetrate beyond the skin layer to fatty layers below, and are sometimes classified as 5th, 6th and 7th degree burns, depending on whether, and to what extent, muscle or bone are affected.

Long term, casualties may be much more prone to internal bleeding and infection due to bone marrow damage, so should be extra careful to protect against physical damage, open wounds, and avoid situations where there is potential for infection.

Full thickness burns will generally need medical attention and skin grafts to heal. Superficial burns, and often partial thickness burns can heal after first aid treatment.

For both radiation burns and heat burns, you need to be careful not to cause further injury, as burned skin is very fragile. For severe or painful burns, clothing should be carefully cut off with a tough scissors or with a knife (cutting upwards away from the casualty's skin), rather than pulling the clothes off, over delicate skin. Burns should be carefully washed. If possible, keep burned area under running water for 20 minutes, or if running water is not available, submerge in cold water and agitate water regularly, to keep the burned area cool. Superficial burns should be moisturised with an unscented moisturiser, (assuming neither the casualty nor the ointment is contaminated with fallout). If possible, use moisturiser that contains shea butter or aloe vera. If moisturised regularly, the wound will heal faster and with less scarring. For more severe burns, apply antiseptic cream or moisturiser to the wound regularly once it has healed over and is no longer draining fluid. If these are not available, extra virgin olive oil can be used, which has beneficial antioxidant and anti-microbial properties. Apply a dressing to the wound and change it daily Casualties should wear loose-fitting clothes or bandages over the burn injury. If the casualty is in a lot of pain, provide oral painkillers, if available. The initial pain of burns may be relieved by cutting strips of cling film and placing it over the burns in layers. Avoid wrapping the cling film around the burn like a bandage, as this will cause pressure to be applied if the injured area swells, which will become painful. Cling film is ideal for covering burns, as it will not stick to the wound and can easily be peeled off when further treatment is applied. If the casualty has facial burns, it will be impractical to apply cling film, so apply burn gel, if available. During the healing process and for up to a year afterwards, protect the wound from direct sunlight, as the healing skin will be highly susceptible to damage by ultra-violet light.

For partial thickness burns, if medical help is not available, it may be necessary to carefully remove dead tissue from the wound, together with regularly application of antiseptic cream and daily dressing changes.

Uranium Exposure

If the casualty has been exposed specifically to high levels of uranium, through inhalation, ingestion or skin contact, then sodium bicarbonate can be taken to alkalise that casualty's urine to prevent deposition of uranium in the kidneys. Typically, treatment consists of 2 x sodium bicarbonate tablets taken every four hours for a three-day period, whilst monitoring the pH (acidity) of the urine and adjusting treatment as necessary to achieve a urine pH of 8-9 throughout that period. This can be achieved using a calibrated pH meter, or disposable pH testing strips, which are commonly available for swimming pool/hot tub testing or from a pharmacy.

Radioactive Caesium and Thallium Exposure

The ingestion of radioactive elements Caesium and Thallium can be treated by taking Prussian Blue tablets (NB **not** the artist dye). This is not an 'over the counter' drug and can only be prescribed by a physician. Prussian Blue itself is not readily absorbed through the gut wall, so it tends to pass straight through the body. It also has a great affinity for heavy metal ions, and it causes the ions to stick to it, preventing them otherwise being absorbed through the gut wall and ending up in the bloodstream. This process is known as chelation. By administering Prussian Blue, the biological half-life of caesium (how long it takes for the concentration within the body to reduce by half), reduces from 110 days to 30 days, and that of thallium from 8 days to 3 days. This is basically the time taken for the concentration inside the body to reduce by 50%. Prussian blue causes no serious side-effects, but some people may experience stomach pains and constipation. When taking Prussian Blue, stools will appear blue for a few days, due to the intense colour of the product.

Suturing

After decontamination of a wound is complete, the wound may need to be closed. Small wounds that are not too deep, and are not infectious can be held closed manually, and superglue applied along the line of the wound to hold it closed. If possible, use some kind of mechanical binding to help hold the edges of the wound in place, such as a dressing and a bandage. Be mindful that there is a risk of scarring and infection.

If the wound needs suturing (stitching), then it definitely needs medical attention, so only attempt to stitch something up yourself if there is no possibility of medical emergency service care. You can buy suturing practice kits online, which will have the required equipment and a layered silicone model for practising on. Doing so will allow you to hone your skills on a skin-like substitute. You can also find videos online showing you how to do this. I have provided a short description below:

Firstly, attempt to get professional medical help. If this is not possible, and the wound is sufficiently serious that it must be stitched, then, and only then should you attempt it yourself. You should only do this if:

1) There is no possibility of professional medical care
2) You are confident in your ability to perform this surgery
3) You have practised, and your practice was successful
4) You have the consent of the casualty

Process for Suturing:

Obtain sterile tools and sterile stitching material.

Ideally you need the following surgical instruments: a needle driver (a surgical pair of pliers used to grip the needle and guide it through the skin), tissue forceps – used to manipulate the damaged tissue into the right position for stitching, scissors – for cutting the thread, and a curved needle and thread for forming the suture. All tools, thread and needle should be sterile. Some suturing kits contain a sealed pack of sterile instruments which will remain sterile until the pack is opened. If your instruments are not sealed inside a sterile pack, you will need to sterilise everything before use. If you do not do this, there is a very strong chance of introducing infection into the wound, which could prove serious or fatal. If you are purchasing suturing equipment, make sure the needle/thread is sterile packed rather than buying non-sterile needle/thread packs intended for practice only.

To sterilise the equipment, there are a few options: Soak for 30minutes in a 0.1% sodium hypochlorite solution (bleach). Rinse off the bleach with sterile water afterwards. Alternatively, soak everything in 70-80% isopropyl alcohol for

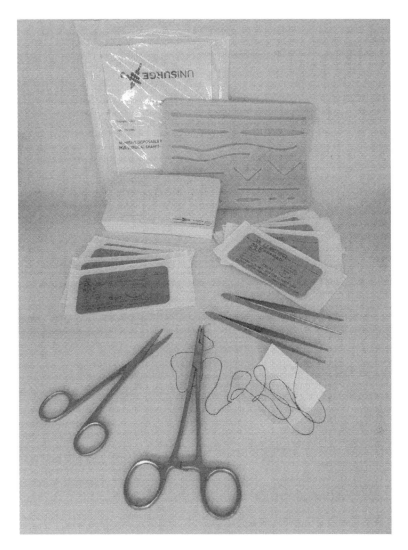

Figure 17: A suturing kit purchased online. The silicone block at the back contains simulated wounds for practicing on. In front of this is a semi-realistic skin replica, into which you can create your own wounds for suturing. The blue packs contain sterile threads attached to suturing needles, one of which is clamped in the jaws of the needle driver in the foreground. In this kit, the instruments are supplied sterile packed, although they have been removed from their sterile packaging for the photograph.

30 minutes or wrap the instruments in aluminium foil and place in water inside a pressure cooker. (Also known as an autoclave). Heat until it starts venting steam and continue to do so for 20-30minutes. Let the pressure cooker cool down and the tools should be sterile when they are removed. Pressure cooker sterilisation works on the basis that it heats to around 121C and runs at a pressure of about 1bar (15psi) over atmospheric pressure, which is able to kill most germs – even those that can survive in boiling water. When you remove the wrapped tools, press the aluminium folds tightly to ensure the tools are sealed inside, to prevent them becoming contaminated with microbes and spores from the air. If you plan to do this regularly, you can buy special bags to seal the tools in, which, when you remove them from the pressure cooker, will indicate that they have been correctly sterilised. As long as the tools are not removed from these bags, they will remain sterile.

If you are unable to achieve any of these methods, you can simply boil the tools in a pan of water for 20 minutes. If you cannot do this, you can simply scrub the tools with soap and rinse with tap water. It will not be as effective and will lead to higher infection rates but will be better than nothing at all. If you have no suitable suturing equipment, you could even use a regular needle and thread, but it will make infection much more likely, and it will be difficult to achieve. Suturing needles are ideally curved to make the job easier.

Making the stitches

The simplest method is to make a series of individual stitches along the length of the wound, drawing the two sides together. This is known as an *interrupted suture* because each stitch is totally separate from the other stitches.

You will need to start by scrubbing your hands. Clean them by scrubbing with soap and water. Make sure you clean all parts of your hand and pay particular attention to your nails, between fingers and creases in your skin. Dry your hands in a clean, unused towel. Put on disposable latex or nitrile gloves. Ideally wear one pair on top of a second pair. The reason for this is that the thin material that surgical gloves are made from tends to tear relatively easily. If you have a second pair of gloves underneath, even if you do get a tear, you should still protect yourself and your casualty, as there will be an intact glove underneath. Take extreme care to avoid puncturing yourself with the suturing needle. If you do so, you may potentially contaminate yourself with any blood-borne disease and likewise, if you continue to use the needle on the casualty, you may endanger them too. For this reason, use the tools to manipulate the needle and do not force it.

Before you start stitching, ensure that the wound is thoroughly cleaned and flushed with tap water or saline solution. Ensure that any foreign material is removed from within the wound and that the wound is fully decontaminated. Remove as much blood from the wound as possible, so that you will be able to see what you are doing. However, if the wound has stopped bleeding, take care not to restart the bleed. Using the radiation monitor, check the wound carefully for radioactive contamination, and only proceed when the wound is completely contamination free. Whereas it is generally advised NOT to use antiseptic cream on larger wounds, if you are dealing with a wound in a battlefield or survival situation, where professional medical care is not available, then it is a good idea to apply antiseptic cream to the wound.

Warn the casualty that this will be painful. Doctors will use anaesthetics to numb the area before proceeding. If you have this to hand, and you're qualified to do so, then go ahead and use it, but unless you are a paramedic or you are in a hospital or ambulance, the chances are that you will not have this luxury.

Thread the needle, and then using the needle driver (medical pliers), grip the needle in the middle, The needle driver will usually have a lock on it – when you squeeze the finger grips together, they will latch onto each other, so you can manoeuvre the needle without having to maintain the force.

Align both sides of the wound, and start at one end, approximately 5-10mm along the end of the wound. Insert the needle through the skin on one side of the wound – approximately 1cm away from the cut itself, and push the needle down through the skin, rotating your hand which will bring the curved needle back up through the skin on the opposite side of the wound, about the same distance away. Aim to penetrate the skin layer fully, but making sure that the needle does not penetrate the fat layer which is immediately below the skin. As you push the needle through, the needle driver will end up close to the skin. At this point, unlock it, and use it to grab the needle near its point, where it has come out on the opposite side of the wound. Using the needle driver, pull the needle on through, and keep pulling thread through until there is only around 5cm of thread remaining. Release the needle driver so it no long grips anything and then with the grips close together, using your free hand, wrap two turns loosely around the nose of the

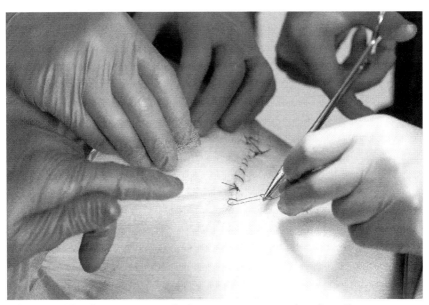

Figure 18: A surgeon stitches up a wound using the interrupted method of suturing.

needle driver. Then, taking care not to drop these loops off, grip the free end of the short (5cm) section of thread on the other side of the wound with the needle driver, and allow the loops to slip off the nose of the needle driver, over the free end of the thread. Do not release the needle driver pressure, and pull the thread with the other hand, to create a surgeon's knot. Tighten just enough that the two sides of the wound touch. Then create a second 'throw' of the knot by repeating the process from the opposite direction with a single loop over the nose of the needle driver, and then again to create a third 'throw', in the opposite direction with a single loop. Pull the knot to one side or other of the wound, so that it does not sit directly over the top of the wound. Cut both ends of the thread so that there is around 5mm protruding from the knot.

Repeat the above to make another stitch approximately 5mm further long the wound. Continue with stitches until the two sides of the wound are pulled completely back together.

Broken Bones

A whole range of injuries can be described as broken bones – e.g., relatively minor cracks in bones, to semi-severed limbs or bones protruding from flesh. These types of injury can easily occur as a result of exposure to blast, or due to falling masonry or falls from height.

If a bone is protruding (an open fracture), firstly treat any bleeding by applying pressure around the fracture, but not directly over the protruding bone. Clean the wound, and carefully apply a sterile dressing, or if this is not available, any clean lint-free cloth. Advise casualty to avoid putting weight on the broken bone. Seek immediate medical attention if available.

Splints may be used to support a broken bone in order to prevent movement of the break, which will be painful and may make the injury worse. Use any stiff material to brace the length of the broken limb across the break and strap it to the limb both above and below the break. Use whatever you have to hand. Where nothing else is available, you can use a rolled-up newspaper or even two rolled up newspapers placed on either side of the arm - to splint an arm fracture. Wrap bandage around the newspapers and arm to hold them together, or if unavailable, wrap and tie any cloth around the arm, both above and below the break, or even use cling film. One valuable addition to your First Aid kit, is an aluminium-centred flexible foam splint (e.g., 'Sam Splint' manufactured by 3M) for forming stiff splints 'in the field'. These can be used to quickly and securely stabilise a huge variety of bone injuries. *Figure 19* shows some examples of how these can be used. Note that a bandage should be wrapped around the splinted limb to secure, but this is omitted from the photographs for reasons of clarity. It should be noted that the splint should be curved in profile for greater strength, and the shape should be formed around a volunteer's limb, rather than the casualty's – to avoid exacerbating the casualty's injury in the process.

Do not allow the casualty to eat or drink if they are likely to need surgery - assuming you are able to get them to professional medical care.

For suspected broken pelvis, the casualty should not be moved, as this can cause complications, including severe internal bleeding. If the casualty is lying down with toes pointing sideways in opposite directions, this is an indication of a broken pelvis. If the casualty must be moved, then you should tie the legs, together at the feet, carefully bringing them together so that the toes point upwards and stabilise in this position. The legs should be immobilised, and the casualty transported on a board of some kind, after sliding them onto the board without moving their legs relative to the rest of their body.

Treating Medical Shock

There are all kinds of scenarios where casualties can go into medical shock. Even relatively minor injuries can cause some casualties to go into shock. Shock causes a reduction in blood flow, which can put casualties in a critical condition, so it is important to recognise the symptoms and try to treat it.

Symptoms of shock:

> Pale/bluish skin (or greyish - for darker-skin tones)
> Cool, sweaty skin
> Rapid, shallow breathing
> Rapid pulse rate
> Weakness
> Being somewhat disconnected from reality
> Becoming anxious or agitated
> Dizziness or fainting
> Feeling or being sick
> Becoming unresponsive

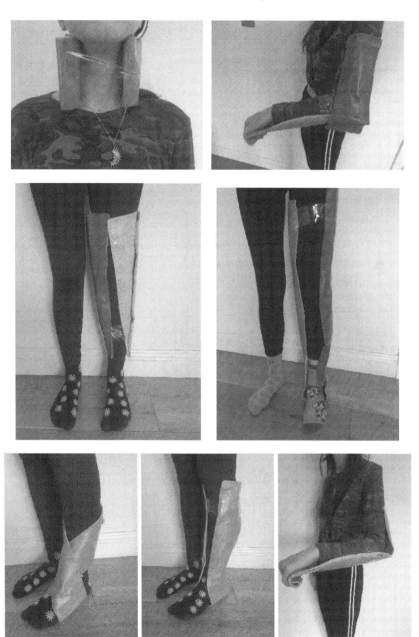

Figure 19: A variety of examples of single and double applications of flexible splints: Top row - a) neck brace, b) upper and lower arm brace. Middle row - c) knee brace, d) whole leg brace. Bottom row: e) ankle brace, f) ankle and lower leg brace, g) lower arm brace.

Treatment

Firstly, treat any bleeding that might be responsible for the shock. Assuming that the casualty is breathing and has a pulse, reposition them so that they are in a lying position with their legs raised up substantially. Loosen any tight-fitting clothing around the waist and neck to remove any restrictions to blood flow. The casualty will potentially lose body temperature, so cover with a blanket (or if that is not available, use additional clothing, cushions etc. – whatever you have access to). Fear and pain will make the situation worse, so throughout the process, reassure the casualty, try to keep them calm, and talk to them in a calm manner. Even if your casualty appears to be unresponsive, continue to do so, as hearing is the last sense to go as a casualty loses awareness of their surroundings. If the casualty does become unresponsive, prepare to administer CPR.

Minor cuts and abrasions

Any broken skin should be thoroughly washed to remove contamination, and the wound dressed with a dressing and a bandage. If the casualty has badly damaged or burnt skin, avoid using sticky tape or sticking plasters, as it may cause pain, or rip the skin off. Instead, wrap bandage right around and only use tape to stick it to itself.

12 PROTECTING AGAINST IODINE-131 EXPOSURE

One of the radioactive isotopes created during a nuclear emergency is Iodine-131. This is highly radioactive, and has a short half-life of 8days – meaning that after 8days, half will have decayed away, and after another 8days it will decay to a quarter of the original level etc. This is in addition to the reduction in concentration caused by wind dispersion and diffusion. The problem with this specific isotope is that it is readily absorbed by the lungs or through the gut and concentrated in the thyroid gland, where it can ultimately lead to thyroid cancers. Those most at risk are young children and to a lesser extent adults under the age of 40. Adults over 40 are considered at lower risk.

Many people assume that iodine is present in the form of a vapour, however, iodine has a low boiling point (184°C) and a low vapour pressure (0.027kPa) compared to air (101kPa), so it will not be present significantly as a gas. The descent of radioactive fallout is due to gravity, causing the small particles of dust to fall. These dust particles are created as the fireball cools. As iodine gas would not fall to the ground under the effects of gravity, it will not be strongly represented in the fallout. However, in chemical terms, iodine is relatively reactive, especially given the temperatures and pressure of the aftermath of the nuclear explosion. Therefore, iodine will generally react with gases in the air, with vaporised components of the nuclear bomb itself, and with material dragged up from the ground, forming both inorganic and organic chemicals – a few may produce gaseous products, but most reactions will result in the production of solids or liquids. These iodine-containing compounds will fall to the ground and quickly find their way into water courses and will be taken up by plant life on the ground. Elemental iodine is also adsorbed onto small particles of fallout and can make its way down to ground level in this manner. The radioactive iodine thus enters the food chain and is especially present in milk. The main route for exposure therefore is not through inhalation of iodine gas, but through the inhalation of fallout particles which have reacted with or adsorbed the radioactive iodine, or though consumption of contaminated water or food.

The effects of iodine-131 can be alleviated by ensuring the body has a good supply of non-radioactive iodine, so that the thyroid does not seek to absorb more – in this situation, the radioactive iodine is secreted from the body and causes relatively minimal damage. In order to achieve this, it is necessary to take potassium iodide supplements as soon as possible after the nuclear emergency takes place, (ideally starting 24 hours before exposure, or within 4 hours of detonation) and continue for at least 8 days. The dose will depend on the age of the person, see table below:

Exposed Person	Daily Dose
Adults 18-40 or teenagers of 70kg or more	130mg (0.13g)
Children 12-18 (under 70kg)	65mg (0.065g)
Children aged 3-12	65mg (0.065g)
Children of 1mont - 3years	32mg (0.032g)
Babies under 1month old	16mg (0.016g)

Breastfeeding mothers should take potassium iodide supplements as radioactive iodine can be passed on from mother to baby. If it is possible to safely feed baby with alternative food source until mother is removed from exposed location, that is preferable.

If the potassium iodide supplement upsets your stomach, take it after meals or with food or milk unless otherwise instructed by a doctor. If stomach upset continues, consult a doctor.

Adults over 40 years old are less likely to suffer from thyroid cancers, and more likely to suffer from complications arising from high doses of iodine. However, it may still be useful, depending on exposure levels – so consult public health official or doctor familiar with the incident.

Only take the iodine supplements whilst a risk of exposure to radioactive iodine exists. Do not take longer than necessary.

Do not use if the potassium iodide has turned brownish yellow.

Take potassium iodide in a full glass (8 ounces) of water or in fruit juice, milk, or broth to improve the taste and lessen stomach upset. Potassium iodide has an unpleasant, bitter taste, and studies have been carried out to identify which drinks it can be mixed with, in order to mask the bitter flavour. Results demonstrated that fruit syrup was the best at masking the bitterness, but orange juice and cola gave acceptable results. Water and unflavoured milk were deemed the least effective at masking the flavour. Be sure to drink all the liquid to get the full dose of medicine.

Potassium iodide is not a treatment and cannot reverse damage already done to the thyroid, it lessens the effect of exposure, but does not give 100% protection.

13 PROTECTION AGAINST ELECTROMAGNETIC PULSE (EMP)

Electromagnetic pulses (EMPs) can be generated by various natural and man-made phenomenon. However, the EMP created by a nuclear explosion ('NEMP') is far more intense. Although the NEMP is unlikely to have any direct health consequences, it has the potential to cause serious problems for electrical equipment and can damage electronics. According to simulations run by UK Utilities, it can also cause blackouts, although it is unlikely to cause the whole of the UK Electrical grid to collapse. Some nuclear bombs are designed so that the EMP is actually the main attack feature of the weapon – by detonating in the stratosphere, they create a much more intense EMP than if detonated at ground level.

Electrical items likely to survive an EMP.

- Cars – despite the large amount of electronics in a modern car, it is thought that a good percentage will survive an EMP, at least remaining drivable. This is because the electronics in a car are relatively well shielded. As a general rule, the simpler the technology, the more likely a car is to survive, so older vehicles without ECUs (electronic Control Units) are almost certainly going to be unaffected.
- Batteries – Single-use batteries and cells, lead-acid batteries, and rechargeable NiCd (Nickel Cadmium) or/ NiMH (Nickel Metal Hydride batteries are all generally unaffected by EMP. Lithium-ion cells should be unaffected, but batteries made up of multiple lithium-ion cells generally incorporate some kind of battery-management electronics, which is susceptible to EMP damage.
- Power Tools – simple power tools without motor speed control are likely to survive. Those with speed controllers are more susceptible.
- Solar Panels – these should generally survive EMP, although bypass diodes may fail and need to be removed, and those with in-built or associated micro-inverters for maximising output may be adversely affected.

Items that <u>may</u> be damaged or destroyed by a NEMP:

- The electrical power grid.
- Anything connected to a mains charger or electrical outlet.
- Computers.
- Electronic controls (if present) in petrol/diesel generators.
- Consumer Electronics.
- Heart Pacemakers.
- LED flashlights.
- Radios and Television Sets – especially if connected to external antennae.
- Mobile phones.

I should stress that items <u>not</u> connected to long wires (such as power cables or antennae) may survive a NEMP, even if they are in the list above, but there is also some chance that they may fail.

It is possible to design installed systems to protect against EMP – for example by encasing any cables inside grounded metal conduit, fully enclosing equipment in earthed metal cabinets and with the addition of transient voltage suppressor diodes and NEMP power-line filters.

Surviving EMP

EMPs affect electrical conductors, metalwork and electronics. It is generally not harmful to humans except for a risk of burns from large sparks near the source of an EMP, but you are unlikely to survive the nuclear radiation, the blast or the heat at this distance. There is usually very little warning of impending EMP, but if you have a warning, ensure you are a decent distance from any metalwork or electrical wiring, as it can cause a lightning-like discharge.

One of the main concerns about EMP is that it can damage electronic devices such as radios and radiation counters which may be of critical importance to survival. To deal with this problem, you need to put any critical device inside a Faraday Shield – this is basically a hollow metal structure which completely surrounds the electronic device, whilst being electrically isolated from it. When an electric field meets the conductor of a Faraday shield, it causes the conductor to become charged. This charge remains on the outside of the shield and the electric charge inside an electrical conductor remains zero. This is only true for fully enclosed shields. If any gaps exist, then an electric field may enter. Expensive Faraday shielding boxes can be purchased, but it is very easy to make your own using tin foil. Tin foil is the source of many paranoia jokes, but in fact several layers of tin foil create a very effective Faraday shield. To do this, take your electronic device, and wrap it a couple of times in paper (which will provide the insulation), then wrap it with 5 layers of aluminium foil. Make sure the foil completely surrounds the object with 5 layers on every side.

Once it is completely wrapped, use cling film or tape to hold the foil down to stop it peeling off.

You do not need to worry about batteries/cells – these will not be damaged by EMP, with the exception of lithium-ion batteries if they have built-in battery management systems.

When No One's Coming

14 EXPANDING YOUR EMERGENCY SUPPLIES

Beyond the essential emergency equipment, decontamination products and food/drink, there are many other things that can be stockpiled to make life easier or safer. Obviously, there are a few considerations here: How much money do you want to invest in preparing for something that hopefully will never happen? How much space do you have for storage? How many eventualities do you want to prepare for? And so on,

First Aid – You will need a decent first aid kit, with extended capabilities to cope in situations where Emergency Services may not be available. See Chapter 11 for suggested contents.

Medication – Some individuals will depend heavily on medication for survival or for their quality of life. These life-preserving drugs would be as essential a part of their emergency store as food or water. Even if the medication you rely on is not essential for life, it is worth stocking up, whether it is prescribed medication, or 'over the counter' (unprescribed medication). You may wish to stock up on painkillers, antiseptic creams, hand sanitiser, antiseptic wipes, diarrhoea medication, laxatives, as part of your emergency First Aid response.

Vitamin Tablets – In the event of a nationwide or worldwide nuclear event, it is likely that fresh foods will be the first to disappear from the supply chain. In the event that you are living off dried/preserved foods for long periods of time, it is likely that you will not get the daily recommended dose of vitamins, so stocking up with 6-12months supply of vitamins for each member of your family is a good investment.

Personal Protective Equipment ('PPE') – ideally everyone in your family needs to have masks. Ideally full-face masks with P3 Radiation filters, but if you are confined by storage space or finances, then even the disposable surgical masks are better than nothing. Even though these are deemed to be single-use disposable items, they can be reused a number of times if necessary, and if you are careful with them. However, if they are exposed to significant contamination, they should be safely disposed of and replaced. Therefore, you will need to consider storing 10-20 times as many disposable masks as there are people in your family. Eye protection is also a good idea for every person, to limit the exposure of your eyes if it is essential to travel through contaminated locations.

Decontamination Consumables - e.g., disposable paper towels, bulk soap, wet wipes etc.

Disability aids – Include a spare pair of glasses for reading or for long-distance vision, contact lenses and cleaning solutions, hearing aid and spare batteries, walking stick.

Baby products – anything needed for your baby or toddler. Things like nappies (diapers) – both disposable and reusable, baby food or dried baby milk, feeding bottles/spoons, wet wipes, rash cream and baby medicine

Pet food/cat litter – depending on the size and quantity of pets, they may consume greater or lesser amounts of food. You should, as an owner, already have a good idea of the amount of food that they consume in a week or a month, so you can use this to gauge how much you want to store. It is worth noting that dried pet food is often available in much bigger bags that wet food, stores well, and is well-priced. You do need to consider the extra water requirements of pets, especially if they are eating dried food, and increase the amount stored to take this into account. It is worth considering pet carriers for transportation purposes, if you need to evacuate. You will need to keep your pet indoors, perhaps for an extended period – so ensure you have adequate supplies of litter, and pet toys to avoid frustration and boredom.

Tools/Rope/String – It is impossible to know what task you may need to complete in an emergency, but you will likely need tools to do so, or to do so efficiently. Tools such as shovels, spades, saws (bow saw and hand/jack saw),axe, hammers (and nails), screwdrivers and screws, knives, pliers, multi-tool, wrenches, chain saws, camping stoves and gas bottles, radiation monitors and protective equipment, portable solar or hand-crank chargers, radios (two way and broadcast receiver), fire-lighting equipment, wide tape, pallet wrap cling film, tarpaulins or polythene sheeting etc., may all prove useful. Heavy duty rope and lightweight string are invaluable for all sorts of makeshift structures or facilities. Manual tools are potentially more useful than electrically powered or petrol-powered tools, as there may be difficulties in finding a source of electricity or fuel after a nuclear event.

Cash/Tradeable Resources – Cash has debatable value in an emergency. In times of hardship, cash is king, and is of more use than currency on a bank statement, which may be entirely inaccessible in an emergency, or may have access restricted by the bank or by the Government. However, in times of crisis, you may find that cash has a much lower value than tradeable items such as food, medication or equipment. It is worth considering having sufficient tradeable resources rather than a lot of cash.

Clothing/Bedding – for each person, have a few changes of clothes including plenty of warm layers, plus sufficient bedding and blankets to cope with freezing temperatures. If possible, include a sleeping bag for each person, in case it is necessary to sleep in the open. Ensure that sturdy shoes/walking boots are available for each member of the family and adequate provision for carrying babies and toddlers (if appropriate), potentially for long distances.

Sanitary/cleaning products – toilet rolls, soap, feminine hygiene products, razors, bleach, toilet cleaner, disposable cleaning cloths, detergents, liquid soap for washing dishes etc.

Important Documents – driving licence, passport and birth certificate for each member of the household, as well as bank cards, credit cards, insurance documents etc.

Fire-lighting/extinguishing – e.g. matches, lighter, magnifying glass, flints, kindling as well as a store of firewood if possible. You may also need fire extinguishers and fire blanket, although these are only suitable for controlling a relatively small fire. Also, a smoke alarm and Carbon Monoxide alarm may be useful.

Writing equipment – Notepads, pens. Pencils and permanent markers, chalks, line-marking spray paint.

Items of Faith: During a crisis, even individuals who are not particularly devout, may turn to their faith for reassurance. Therefore, you may wish to include religious items relevant to your faith such as: prayer books, religious texts, hymn book, prayer/dhikr beads, crucifix, prayer mat, compass, Skull cap/Kufi, Menorah, Wash bowl, etc.

Cooking Facilities – either a camp stove with spare gas cylinders, or kit for cooking on an open fire outdoors. Make sure you have a few suitable pans and include lids, as these will increase efficiency – thus making bottled gas go further and will also decrease cooking time. Also include sufficient unbreakable cups and plates as well as knives, forks, spoons and a couple of can openers.

Items to relieve boredom – You may find yourself having to spend many hours or even many days in confined space with nothing to do. Therefore, pack books, playing cards, musical instruments, games and puzzles, especially if you have children to entertain. It is also worth storing a good book of card games – so you can learn new card games when you get bored of the ones you already know.

15 FOOD

In an emergency, food shortages are almost inevitable. During the Covid pandemic, we witnessed long durations of food shortages, bare shelves in supermarkets and rationing of popular food products by the supermarkets themselves ('no more than three of any one item' was a popular phrase at the time). Although we did not run out of food, and nobody starved as a result of the pandemic, there were acute shortages of some foods and some non-food products, and it gave the public a taste of what it may be like to live with food scarcity. Panic-buying became a reality, and this compounded problems – especially for long-life foods such as pasta and flour. The government and supermarkets were trying to reassure people that there would be enough for everyone if they stopped panic-buying, but in reality, those who did not stock up were those who went without.

Taking the UK as an example, it has a much larger population than it can support. Currently, the UK can produce around 52% of the food it requires and imports the other 48%. So, if we were suddenly unable to import food (which is fairly likely in the event of a nuclear war), we would only be able to feed around 35million people out of our population of 67 million. Of course, in the event of a nuclear war, we would most likely not be able to import the herbicides, pesticides and fertilisers which allow us to maximise food production. In fact without this intensive farming, food yields would likely reduce by 20-40% meaning in the worst case scenario, we would only be able to feed 21 million people, and that does not include any reductions due to interference caused by war – e.g., specific targeting of farms with conventional weapons, the inability of farmers to produce food due to fear, lack of fuel or contamination of food by radioactivity, deaths of animals due to radioactive poisoning, or failure of crops due to nuclear winter.

In addition, the UK only has around 2 weeks-worth of food in the supply chain – a weakness that was highlighted by the Covid pandemic - so if, for some reason, there was suddenly an incident preventing food imports, there would be bare shelves in the supermarkets within two weeks. During the pandemic, food import and delivery services were prioritised to ease this problem to some extent – an option that is unlikely to be available in the event of a nuclear war. Of course, when you factor in panic-buying, you'll see bare shelves in a matter of days, and this was all too obvious during the covid pandemic. In the event of a major disaster in the UK, individuals would quickly need to adapt to being either self-sufficient in food production (a very hard thing to achieve, especially if subject to wide-spread radioactive contamination) or rely on their own food stores to a greater extent. The same applies for any small country with a large population, or any country with large amounts of land not ideally suited to agriculture.

An average person eats 1-2kg of food per day. The exact amount obviously depends on what the food is, and how much water is contained within the food, but you can calculate roughly how much food you need to store by multiplying the number of people, by the number of days you want to store food for, by 2kg. e.g.,

4 people x 100 days x 2kg = 800kg.

This is a reasonable rule of thumb, but of course, it's not quite as simple in reality: You can survive on less food than you'd actually like to eat, but if you are very cold because you can no longer heat your house, or you no longer live in a house, you will require more food, as you burn more calories simply to maintain body temperature.

When considering the amount of food, you need to store, bear in mind that most foods contain quite a lot of water, so if you can store dried foods, you won't need to store as much, as they take up less space - in addition to generally having longer shelf lives. You will also need to make provisions to provide clean, uncontaminated water to rehydrate dried foods.

When selecting foods to store, the characteristics you are looking for are:

1) Long Shelf-life,
2) Compact foods that do not take a lot of space,
3) Foods that can be stored at room temperature,
4) A variety that provides for a balanced diet.

An Example of a very basic store of staple food for one adult for one year is shown below:

Cereals (e.g., wheat / rye / flour / rice / pasta / barley / oats / millet)	120kg
Powdered Milk	40kg
Dried Maize (Sweetcorn)	120kg
Dried Legumes (soybean, lentils, chickpeas)	60kg
Fats/Oils	10kg
Sugar	20kg
Multivitamin tablets	365 tablets (One per day)

Note that the shelf-life of whole grains is substantially longer than when ground up. For example, whole wheat may be stored in a dry, oxygen-free environment for 30 years, whereas flour may only last 5 years.

A diet based entirely on bulk staple foods would be extremely bland, but by stockpiling sauces, spices, and tinned foods, it can become quite acceptable to eat. For example, a bowl of cooked, plain rice eaten on its own might be quite unappetising. However, even the addition of a small amount of spice to add flavour can make a huge difference. Accept that in a mid to long-term emergency, there will need to be big changes in your diet in order to survive on stockpiled food beyond a few weeks, and that it will not be easy, or even achievable for your diet to be as healthy as usual. However, the more food you can store, and the more effort you make to include variety, the more interesting and palatable your survival diet will be.

Aim to provide the bulk of your calories and nutritional requirements from your staple foods and subsidise these with small amounts of stronger tasting foods to keep it interesting.

For a modest food store, without going to extremes, aim to stockpile foods that have a reasonably long shelf-life and 'rotate' on a regular basis (i.e., use the oldest items and replace them with new). To keep costs down, buy products when they are on offer (e.g., buy-one get-one-free deals), and switch to supermarket 'own-brands' or 'value' ranges to keep costs down (when there is no other food available, it is unlikely you will care too much about the branding!). Keep the food store well-ordered, so it is easy to rotate the oldest items.

Food/Drink items to consider buying for long-term stockpiles:

Instant Coffee / Tea bags / Hot Chocolate
Powdered Milk / UHT milk
Long-life Juices / Canned or Bottled soft drinks / Bottled water
Fruit Squash Concentrates
(Alcoholic drinks)

Cereals
Porridge Oats/Rolled Oats/Museli
Rice (Short Grain/Long grain/Basmati
Flour
Sugar
Dried legumes (peas / beans / lentils)
Pasta
Instant mashed potato

Vegetable Oil
Salt / Pepper/ Spices / Stock cubes
Vanilla Extract
Dried Soya / Textured Vegetable Protein

Pasta Sauces / pesto / tomato puree

Tinned Fruit / Dried Fruit / Nuts
Tinned meals / beans / spaghetti / soups / tomatoes / vegetables
Tinned Fish/Meat

Biscuits

Honey
Jams
Savoury Spreads
Peanut Butter
Corn Syrup
Chocolate spread

Margarine / Butter (but only short-term)

Chocolate bars (dark chocolate keeps longer)
Protein bars
Granola

Condensed Milk
Coconut Milk
Gravy Granules
Instant Custard Powder
Soy Sauce
Ketchup

Packet (powdered) Soups

Dried yeast (for baking)
Vinegars (apple, malt, white)
Baking soda

Jelly/Jell-O mixes

Freeze-dried foods (for exceptionally long shelf life)

Obviously, you can vary this list according to your dietary preferences.

When No One's Coming

16 FOOD STORAGE AND SHELF LIFE

All food has a shelf-life. Some food becomes harmful to eat, or at least distasteful, due to bacterial contamination. Some foods simply reduce in quality – they may become dry and stale, or lose their crispness, but are still perfectly safe to eat. Some foods show very little noticeable deterioration over many years, although the nutritional value may reduce slightly.

There are several agents of degradation, which cause food to deteriorate over time. The first is humidity – the presence of higher concentrations of moisture in the food enables certain chemical degradation reactions to take place. Also, as the humidity rises, it can provide a habitable environment for certain insects and bacteria, which may result in massively higher degradation rates when the humidity increases over a certain level. As the chemical reactions involved are oxidation reactions, the presence of higher concentrations of oxygen speeds up the process. Oxygen is also a necessity for most living organisms, so the presence of oxygen will increase the probability of degradation due to insects, bacteria or vermin. Most chemical reactions will double in speed for every 10C increase in temperature (some biological reactions, and enzyme reactions do not follow this rule) – which is why storing food in the refrigerator rather than at room temperature will decrease the rate of degradation four-fold. Storing food in a freezer tends to slow most reactions lower still, not only because the temperature is that much lower, but also because any liquid water is turned into solid water (ice), making any water-based reactions much slower as the reacting molecules are no longer able to move freely throughout the medium. Light also has a degrading effect on foods, due to the damage caused to molecular bonds by ultraviolet light. By combining several of these effects at the same time, degradation rates will simply accelerate.

Biological Agents of Degradation

The simplest form of biological degradation is enzymes. These are special proteins produced by living organisms which, by forcing reacting molecules into close proximity to each other, can massively speed up the rate of reaction. This is known as catalysis, and the enzymes are a type of catalyst. There are many different types of enzymes in nature, but the ones of particular interest to us, in trying to preserve our food, are the degradation enzymes. Some of these are even present in the food we are trying to preserve.

The effect of enzymes can be reduced or eliminated entirely either by subtly changing the shape of the enzyme (e.g., by changing their environment slightly – such as changing the acidity/alkalinity of the liquid they are in, or by heating them to a temperature of 55°C or higher.

Another biological agent of degradation is Yeast. Yeasts are single-celled organisms, similar to plants, but without the ability to photosynthesise. Yeasts like to grow on sugary food and are naturally present in nature on the skins of fruit, often causing a bloom across the fruit skin. Yeasts cannot generally tolerate sugar concentrations over 40% and do not like low temperatures, high temperatures (over 70°C), or high levels of acidity. Yeast degradation normally affects only the taste of foods and does not usually produce any toxic effect. Yeasts have the somewhat unusual characteristic of being able to reproduce and grow in the absence of oxygen ('anaerobic' conditions).

Moulds are another agent of degradation. They are similar to fungi and reproduce by filling the air with billions of microscopic spores, which land on food the moment it is exposed to the air. Moulds are a type of fungi, but do not grow a large fruiting body, like the mushrooms we are familiar with. Instead, they grow what looks like a fluffy, hairy mass on the surface of the food, from which spores are released. Moulds usually have an unpleasant taste, although some are actually used to enhance food flavours (such as those employed in blue cheeses). Some moulds, if eaten, can cause illness, although many are relatively harmless. Moulds generally like slightly acidic conditions and require moisture and will tolerate a wide range of temperatures from 0-70°C.

Bacteria are single-celled organisms and are able to reproduce very rapidly under the right conditions. Some bacteria are perfectly harmless and are naturally present in our guts and on our skins, some are used in the production of foods such as yogurts and cheeses. Some bacteria can be harmful if consumed and can cause illness or even death. Bacteria are present everywhere, but they are generally most harmful when they multiply into large numbers. Some bacteria are also able to grow in anaerobic conditions. Low temperatures reduce the rate of bacterial growth, and reducing the humidity will slow or even stop bacterial reproduction.

'Best Before' and 'Use By' Dates

Commercially packaged food generally has a 'best-before' date or a 'use by' date. These have very different meanings. The 'best before' date is basically a date intended to indicate a period of use when the food will taste at its best. It is normally safe to eat food after its 'best before' date...in some cases, a great deal longer than the expiry date given. Food manufacturers must store a representative sample of their food batches for investigation should there ever be a quality issue. The representative sample can be quite large, maybe as high as 2% of the food produced – and it needs to be stored until the date of expiry. So, if manufacturers give a long shelf-life to a product (e.g., 10 years), they may find themselves renting huge warehouses simply to store all these representative samples. As it is generally not a selling point for the food, it is far easier if manufacturers give a much shorter shelf life, as long as it is sufficient to cope with the likely storage time in the manufacturer's warehouses, the seller's warehouses/shops and still give enough time for a sensible storage period in the buyer's kitchen cupboard. For this reason, you will generally see 'best-before' dates giving 6 months – 2-year time frames. However, if you have ever stored foods for much longer, and opened them to see if they are edible, you will probably find there is nothing wrong with them, even 3 or more years after the best-before date.

The UK Food Standards Agency recommends that you use a 'sniff test' to determine if 'best-before' expired food is still edible. If it smells and looks ok, then it's ok to eat. It might not taste as good, but it is unlikely to do you any harm.

'Use-by' dates are somewhat different – they indicate a food needs to be consumed before it succumbs to biological degeneration e.g., bacterial contamination. Often, in these cases, the bacteria aren't obvious from smelling or looking at the product, and the food can make you ill. So, it's best not to take risks with foods that have 'use by' dates. Having said that manufacturers usually apply use-by dates conservatively, for liability reasons. With both use-by and best-before dates, the manufacturer will have assumed the food is stored in line with their recommendations. If not, the food may degrade faster than anticipated.

How to Extend the life of dry Foods.

Dry foods usually have long shelf lives. The degradation of dry food is determined by a number of factors: 1) Temperature 2) Humidity 3) Oxygen levels 4) Biological Attack

Temperature – Natural degradation reactions, which are chemical in nature, will slow as temperature reduces. For every 10C reduction in temperature, the degradation reaction will reduce by 50%. So, storing food in a cool location will improve its shelf life. Of course, cooling below ambient temperature will take energy – something that may be difficult or expensive to do. The best option therefore is to store the food in a cool location in your house, or if you have access to a basement or purpose-built underground food store, even better, as this will likely have a year-round temperature of approximately 10°C – which will double shelf-life compared to storage at 20°C.

2)Humidity – The effect of Microbial degradation and organic degradation (e.g., enzymes) may not be as predictable as you might imagine in response to temperature. However, these mechanisms are generally water-based, and therefore need moisture to operate. As a general rule, the lower the water content, the slower the reaction will be. Therefore, by packing the dried food inside sealed containers, will maintain this low humidity long-term. (Make sure you don't move it from a warm place to a cool place whilst exposed – otherwise condensation will wet it)

3)Oxygen – The rate of natural deterioration of food is strongly linked to oxygen. Food quality reduces as flavours become oxidised and start to deviate from the preferred taste. Therefore, removing the oxygen will dramatically reduce the natural chemical degradation reactions which occur. Removing the oxygen will allow some foods to be stored for up to 30 years. There are various methods for removing oxygen from food storage containers – blocks of dry ice (solid carbon dioxide) can be placed in the bottom of the container before adding the food. As the dry ice sublimes into gaseous carbon dioxide, it displaces the air already inside the container, and replaces it with carbon dioxide. During this process the lid is not sealed, so that excess gases can escape. Once the carbon dioxide has fully sublimed, the lid can be pushed down to seal it. The advantage of this method is that the cold conditions will also help to kill off any germs that might be present, however the disadvantage is the cold will cause moisture in the air to condense into water droplets, which will be left behind and potentially cause an elevated humidity inside the food container after it is sealed. Another option is to flush the container with carbon dioxide or nitrogen from a storage cylinder, although the cost of renting or buying the cylinder may be quite high. The most popular method of removing oxygen in food storage is to use small packeted 'oxygen-absorbers' – which are basically porous bags containing iron powder with salt as an activator. The iron oxidises in the presence of moisture and oxygen. The salt catalyses the reaction – causing it to proceed faster. These oxygen absorbers are usually rated by the amount of oxygen they can absorb – you can estimate the volume of gas in the storage container, and given that, for round figures, oxygen makes up 1/5 of the air, you can calculate how much oxygen needs removing. Also, by over-specifying, it allows for the absorption of any oxygen that may find its way into the container over time.

It is also worth highlighting a potential danger presented by storing food under low oxygen conditions, for example, canned food, vacuum packed food, food packaged with oxygen absorbers and pickles. Botulism is a rare, life-threatening condition, caused by ingestion of toxins created by the botulism bacteria (Clostridium botulinum and a couple of closely related bacteria). The bacteria are present in the environment, but generally causes no harm. However, when subject to certain conditions, e.g., zero or low oxygen, low sugar, low acidity (less acidic than pH 4.6), water content over 10%, salt concentration of less than 3.5%, storage temperature over 8°C, the spores can germinate and produce toxins which are highly poisonous to the human nervous system. These are the same toxins, in highly diluted formulations, which are used in cosmetic treatments commonly known as 'botox'. Eating improperly preserved foods containing botulism toxins will cause paralysis which spreads from the head down the body, with initial symptoms usually appearing after 12-36 hours, such as fatigue, weakness, vertigo, drooping eyelids, slurred speech, blurred vision and difficulty swallowing or breathing. The disease is serious and requires immediate medical attention and has a fatality rate of up to 10%. The disease is not spread from human to human. Botulism can affect both home-preserved and commercially produced foods.

Precautions to avoid contracting botulism include not eating food from bulging or damaged storage cans and avoiding foul-smelling foods. However, it is important to note that affected food may not appear, smell, or taste bad. Heating food to >80°C will degrade botulism toxins, so adequate heating of foods during preparation is important to minimise risk.

One potential solution to minimise the potential for botulism is to avoid using low oxygen storage conditions and accept the resultant shorter food storage life. Storing dry foods in plastics such as MDPE, that are porous to oxygen, will generally result in food spoilage through other pathways before botulism can occur, and yet still provide fairly long storage life.

4) Biological Attack – Some foods are highly attractive to insects – such as beetles and the larvae of some moths and flies. Sometimes these insects are so small that they are difficult to see, and the only evidence is their waste products present in the food, and debris they leave around food packaging. The Indian Meal Moth (Plodia interpunctella) is a particularly problematic pest found all over the world and can devastate stored grain and grain-derived foods. They are able to chew through cardboard food packaging and also the thin, soft plastic that is often used for packaging foods – so it is essential to store food in hard plastic, metal or glass containers. Unfortunately, the food you purchase is often already infected – which if eaten within a short period will not be noticeably spoiled, as it takes time for the eggs to hatch and go through their reproductive cycles to multiply. Storing food for years or even decades is a completely different matter – and the last thing you want to do is to open a long-term sealed food container in an emergency, to find that the food inside has already been spoiled! For this reason, it is advisable to take steps to prevent and control any potential infestation. Firstly, make a habit of storing food in airtight containers, and do so immediately when the food arrives at

your home. The moths and their larvae cannot easily survive cold temperatures – so placing any food intended for long-term storage in a freezer (-18C or lower) for 4-7 days will kill off any larvae or moths. Some experts recommend a number of freeze/thaw cycles for your packed and sealed foods, which applies more stress to any insect eggs that may be present. It is important that your food is packed in such a way as to ensure no air exchange can take place, before placing it in the freezer. The reason for this is that when warm air hits something cold, it will cause the moisture in the air to condense – like dew on a car windscreen. When you take the food out of the freezer, if it is exposed, it will accumulate considerable amounts of water on any exposed surfaces, as warm air deposits dew on the food until it has warmed up. Moisture is one of the main agents of degradation, so keeping dry food as dry as possible should be one of the main considerations in long-term food storage. Alternatively, food can be heated to 50C for 2 hours, or to 60C for one hour, to achieve the same goal.

Most microbes require oxygen to degrade food. Therefore, by sealing the food inside a container, only a limited amount of oxygen is available for degradation. If you take it one step further - by removing the oxygen, once the food is sealed inside the container, you can almost eliminate this degradation process. This can be done by placing oxygen-absorbers into the container immediately before sealing. In addition, the oxygen-removers will stop the natural chemical degradation reactions from occurring – so the taste, smell, appearance and texture of foods do not degrade over time.

Also, vermin must be considered. There is a high population of vermin in most populated areas, and they have evolved to co-exist with humans. They are very good at locating food from a substantial distance and are able to enter storage rooms through very small spaces – a small mouse, for example, is able to squeeze through a gap only 6mm wide, by dislocating its joints as it does so. Even if the hole is too small, as long as the material is soft enough e.g., plastic or wood, mice and rats are easily able to make it bigger until they can fit though. Once inside, their sharp teeth are easily able to chew through food packaging and containers to access the food. Therefore, it is important to make sure your food is stored in a secure room. If the door is wood, apply aluminium kick plates along the bottom to prevent vermin chewing an access hole. If there is any chance that vermin may get in, make sure you have traps or bait available, which can be used if necessary, however, if you do so, ensure there is no danger of young children or pets encountering them.

Containers such as glass jars and metal cans are ideal as they create a perfect seal against oxygen and moisture, and they are also rodent proof. Some people recommend mylar bags, but gases and moisture can still pass through, to some extent, over the long term, either through imperfect seals, or by diffusion through the plastic itself.

Moisture absorbers and indicators.

The most commonly used moisture absorber is silica gel, which is cheap and easy to obtain. Silica gel is a transparent material – normally supplied as a granular powder or formed into small spheres – which are preferable, as the granular powder can be dusty. Silica gel absorbs moisture from the air over a temperature range of 0-100°C and can therefore be used to reduce the humidity inside a sealed container, or to absorb any humidity that might enter through diffusion or imperfect seals, in order to maintain a low humidity over a longer period. Many silica gel products are impregnated with a chemical indicator – such as cobalt (II) chloride, which changes from a deep blue colour through light blue to pink, to indicate when it has absorbed moisture, When the silica gel has turned pink, you know that it is fully saturated and will no longer be effective at absorbing moisture. This is a very useful feature – you can see at a glance whether the product is still moisture free, and it also allows you to effectively regenerate and reuse the silica gel by placing in an oven on a very low heat (110-150°C), until the colour has changed back to a deep blue colour. Some products use other colour indicators such as phenolphthalein or methyl violet, which turns from orange to green as it absorbs moisture. A less common alternative is ferric ammonium sulfate indicator, which changes from amber (dry) to colourless (hydrated). Currently, this is not classified as toxic, but nevertheless precautions should be taken to avoid direct contact between it and the food. Silica gel is sometimes sold in small plastic capsules with a porous cardboard top. These are useful in that the colour is easily visible, and the silica gel is better isolated from the food product it is added to, but they are less easy to regenerate, as the plastic often melts at a lower temperature than that required to drive off the absorbed water. The indicator colours used with silica gel can be toxic, so it is important to ensure that indicated silica gel is not in direct contact with your food. The orange/green indicator is deemed to be less toxic than the blue/pink indicator. Alternatively, you can use unindicated silica gel and a separate indicator paper – which is usually blotting paper impregnated with a moisture-sensitive colour change chemical. Indicator paper and silica gel should be stored in a moisture-tight container until required, to avoid degradation in storage.

I carried out experiments some years ago, where I sealed silica gel capsules with moisture indicators inside mylar bags, using professional laboratory sealing equipment and stored them in a refrigerator. After 12 months, the moisture indicators had changed colour, indicating that moisture had entered. Of course, there will be variation between batches of packaging materials and between manufacturers - some products may be better than others in this respect. It would be a useful experiment to reproduce using whatever type of container you are using for food storage, to verify how effective your packaging is. If you are using transparent or translucent containers, you will be able to monitor how effective they are as a moisture barrier, throughout storage, without breaking the seal.

It is worth noting that silica gel itself (without any moisture indicator) is not toxic by ingestion, despite the warnings 'Do Not Eat' usually attached to packs of silica gel. It is, however, true that you should not eat it – it is not a food, it is not intended for eating and it could be a choking hazard particularly for young children. You should also take care not to breathe silica gel dust, which, like many dusts, can be harmful to the lungs.

Oxygen Absorbers and indicators

Oxygen-absorbers are generally small porous packets containing iron filings, which react with oxygen in the air to remove it, leaving inert gases nitrogen and argon. The iron filings have salt (sodium chloride) added, to accelerate the oxidation process – basically the iron filings turn rusty.

Oxygen absorber products are often classified according to how much oxygen they can absorb. For example, a 300cc oxygen absorber is able to absorb 300 cubic centimetres of oxygen. As oxygen makes up approximately 20% of our air, this equates to all of the oxygen in 1.5 litres of air. Therefore, if you used a 300cc oxygen absorber in a 1.5 litre container, it would absorb all the oxygen inside that container, even if it was completely empty. If you fill the container with food, it will need to absorb less oxygen as there is less air in the container, and still have plenty of spare capacity to absorb more in the future, should a small amount of oxygen diffuse through over time. Alternatively, you can calculate it to have no spare capacity, in which case you only need to account for the air contained in the headspace above the product and any voids between the particles of packed food. As a typical compromise, most preppers will use 300-500cc of oxygen absorber for every four litres of storage volume. Commercial oxygen absorbers often have activated charcoal too, which is useful in absorbing odours. Some preppers make their own oxygen absorbers, by working finely powdered table salt into fine steel wool. All oxygen absorbers, whether commercially-produced or home-made, must be stored and used in perfectly sealed containers – and this is quite difficult to achieve. If you can achieve this, shelf life of dried foods can be extended up to 30 years in some cases.

It is also possible to buy oxygen indicators. These are coloured capsules, usually pink in the absence of oxygen, and blue or purple in the presence of oxygen, based on a methylene blue dye complex. These are reusable and will change colour within about 30 minutes to indicate oxygen or lack of oxygen.

Storage containers

To achieve the longest shelf-life, it is necessary to choose a container that keeps out humidity and oxygen long-term, as well as being resistant to pests, which rather narrows down the options. Many people may think they have packaged food in oxygen-free environments, but actually oxygen (and potentially moisture) is slowly leaking back in through the plastic itself. Even if it takes a year to do so, the food will only have extended its shelf-life by a year, rather than by a factor of five.

Popular containers for long-term food storage are:

- **Tin Cans–** Strong and durable, with very low permeability, but not cheap and requires specialist canning equipment to seal the cans.

- **Glass Jars** – with tight-fitting metal or glass lids are ideal for long term storage of dry foods. Glass is very strong and durable, impermeable to insects and to oxygen. However, it is fragile, and is likely to be damaged by violence such as earthquakes and bomb blasts.

- **Plastic bottles** – various types of plastic are used for making bottles – PET is the preferred plastic for food storage as it has relatively low permeability to oxygen. If you check the recycling symbol (usually on the base), you can check what any particular bottle is made from. If you search for bottles made from 'PET' (also known as PETE), you will have a cheap container that is relatively impermeable to oxygen and moisture. This type of plastic is often used for bottled water. PET-bottled water is often available in 4 - 5 litre sizes (approximately 1 gallon in Imperial measurements) – often at very low prices. They make ideal containers for storing granular foods that can easily be poured in or out. Just make sure when reusing bottles as food containers, that they are clean and totally dry before filling them with food.

- **Plastic buckets (tubs/containers) with click-seal lids** – These are usually made from PP, MDPE or HDPE, and although oxygen will diffuse through these, they are thick and durable, and can be used for very stable foodstuffs. These are ideal for bulk storage, as they usually come in large sizes (1 – 30litres / 2pints - 7.5gallons), and can be lined with metalised foil bags, sealed with oxygen-absorbers for extending the storage life of foods.

Type of Plastic	O_2 Permeability ($ml/m^2/day$)
LDPE (Low Density Polyethylene)	8000
HDPE (High Density Polyethylene)	3000
PP (polypropylene)	4200
PET (polyethylene terephthalate)	150
EVA (Ethylene vinyl alcohol)	10,000
PVC (Polyvinyl chloride)	2000
PS (Polystyrene)	2500
Metalised mylar	0.5

- *Note: Metalised films have a very low permeability to oxygen, but they are also a LOT thinner than most plastic containers, and the oxygen permeability will be proportional to the thickness.*

Extending shelf-life of other types of food

1) Freezing

Freezing is an effective method of extending the shelf life of foods indefinitely. In practice the taste of frozen food will deteriorate in the long term (2years+), but it will be perfectly safe to eat. However, freezing is energy-intensive – an average large freezer may consume 350-400kWh/year – which at 40p/kWh would cost you £140-160/year. Perhaps even more of an issue would be that there is a high probability that mains electricity will be unavailable or intermittent during a disaster – meaning your frozen food will thaw within a couple of days and need consuming immediately.

2) Freeze-drying

Freeze-drying is an unusual process whereby food is frozen and then desiccated using a vacuum. It is a fairly slow and energy-intensive process, but once the food is freeze-dried, it can be packed in air-tight packaging and stored pretty much indefinitely, at room temperature.

3) Pickling

Pickling is a very old method, particularly suitable for food that tends to spoil quickly such as fruit, vegetables, eggs etc. There is a significant initial outlay in terms of purchasing.

Figure 20: Food storage in PET (PETE) plastic bottles. By kind permission of The Provident Prepper

suitable jars and vinegar, and it is somewhat labour intensive in preparation. Simple home-pickling can preserve food for around 6months, whereas properly prepared pickles can easily last 2 years. Even after 4 or 5years, they are usually perfectly edible – just check the taste to make sure they are not too sour. When pickles spoil, you tend to get obvious mould growth.

4)Canning

Shop-purchased canned goods generally have long shelf life and are perfectly edible a long time after the expiry date. Most canned food can easily be eaten 4 or 5 years after the expiry date, so long as the cans are not rusty, dented or swollen. The exception is only for very acidic foods such as some canned fruit – which can cause deterioration of the can itself. Home-canning can also be done with the right equipment.

5)Smoking

Foods such as meats and cheeses can be preserved by exposing to very slow burning charcoal or sawdust, where it is exposed to thick smoke for periods of 6 to 24hours. Smoking foods usually only slightly extends the shelf-life, and it introduces potentially toxic and carcinogenic chemicals into the food, so is far from an ideal method of preserving. Smoking foods is mainly practised to enhance taste.

6) Salting

Some foods, especially meats, can be preserved by salting, either with normal table salt (sodium chloride), or saltpetre (potassium nitrate) or related chemicals sodium nitrate and sodium nitrite.

The nitrates and nitrites used in salting are deemed to be unhealthy, and have in some cases been linked to cancer, so it is not an advisable method of preserving food. Even table salt is not ideal, as high levels of salt in a diet are detrimental, and people who eat salty foods experience high levels of thirst as a natural reaction to compensate for the elevated salt content of the food. This could be an undesirable characteristic in an emergency situation, where water may be rationed.

Stockpiling Food

Where space and finances are not limited, you may decide to water-proof a basement and turn it into a secure living/sleeping/storage area, where food can be stockpiled in case of disaster. Located in the basement, the cooler conditions are perfect for long term food and water storage. You may decide to dig an outdoor underground nuclear shelter, similar in concept to the air raid shelters of the Second World War, which can also serve as a store for stockpiled food. If you do so, you will need to ensure that the room is not damp and is secure against vermin. Your valuable stockpile will also need to be secured against theft or interference.

Alternatively, if these luxuries are not afforded, you can make do with dedicating kitchen cupboard space, or boxes under the bed, or in the spare room, or wherever you can find the space to store food. Of course, the less space available, the less stockpiled food you can store, but in a real emergency, anything is better than nothing. The less space you have, the more you will need to think about survival and the less about nutrition or tasty/varied meals.

It is important to keep good records of what you have (types of food/drink and quantities), what you need or want to add to your stockpile and what expiry dates you want to use for the foods. If you are repacking food in larger containers such as lidded plastic buckets to protect them from vermin, moisture and to reduce oxygen exchange, you should write on the top and on the side of the container with permanent marker describing the contents, the date it is packed, the manufacturer's best-before date and the expiry date (or rotation date) you assign to that product, so that you can identify the contents easily and quickly without having to open and disturb the contents. Although printed stickers may look neater and be easier to read, over time they can become detached as the glue ages or become damaged or worn. By writing the contents in two different locations, it is easier to identify the contents with less repositioning, and it will still allow you to identify the contents if the writing in one of the locations becomes illegible or damaged for some reason.

Unless you are packing items for very long-term storage (e.g., using oxygen-absorbers and mylar), you should ideally rotate the food every so often, by replacing it with freshly purchased food, and consuming the stored food in your everyday cooking. That way, you will not have food stored for too long. It also makes sense to bulk package food in logical sets – for example items that would make a complete meal when cooked together, or even a full meal package for a set period such as for a family of four for a day, or the food to support a couple for a week.

You could use a shorthand notation to indicate the important details on the top and side of the container e.g.:

---

5 x 800g pasta, bb.02/2025

5 x 500ml bolognaise pasta sauce bb. 04/2024

p. 12/03/2023

r.13/03/2026

The 'p,' indicates the date you packed the food into its protective container.

The 'bb.' indicates the best-before date the manufacturer has assigned to the products.

The 'r,' indicates the latest date you recommend for when it should be rotated for fresh batches.

Note that I have used the UK dating convention here (day/month/year).

Most food and drink manufacturers write their best-before dates in very small writing, often in an obscure part of the packaging. You may wish to make this much more prominent - to make it easier to quickly scan your food store; so for foods that you are not repackaging into protective bulk storage containers (such as plastic buckets), you may wish to use a permanent marker to write the best-before date or your recommended rotation date in large, prominent letters, e.g., on the top of tin cans. You may even want to colour-code different years, so you can scan for red lettering to identify products that need rotating this year, blue for next year, green for the year after etc.

17 WATER

The average person in the USA uses around 300litres per day and in the UK, uses around 150litres of water per day (for all purposes). Actual drinking water accounts for approximately 2litres per day. For the purposes of Emergency Planning, it is generally assumed that each person can get by on less than 5 litres per day for drinking and sanitation.

In an emergency, mains water may be available, or it may not. Even if it is available, it may very quickly become contaminated due to radioactive fallout. The level of contamination may be low enough to carry on using it for washing and decontamination processes, but it may well be hazardous to drink. If you are extremely fortunate, your domestic water may be provided from a borehole – where water is raised from water-bearing rocks and strata, known as aquifers, located deep underground – usually 60-200 metres below the surface. These ground water systems have a water residence time of 50-100 years – so it is likely to run clean for a long time before being contaminated. Very deep aquifers may have water residence times of up to 10,000 years. Do not confuse these with groundwater supplies such as springs or wells, which are generally providing rainwater after it has been filtered through fields and topsoil, in which case it is probably only 1-2 months since it descended as rain – so this water will become contaminated with radioactivity quite soon after a nuclear detonation – but perhaps not as quickly as mains-supplied water from the tap / faucet.

Water is one of the most fundamental requirements for life, so securing a safe supply of drinking water early on is essential. If you have foresight to do so, you may be able to stockpile bottled water in a cool location away from light. Some manufacturers will give a shelf life of 2-3years, but in practice, bottled water can be stored indefinitely if stored under the correct conditions (dark, cool temperatures). Over many years, the water may pick up some taste from the plastic, but it is still perfectly drinkable and safe to do so. You may also store tap water in plastic barrels, which is an easy way to store hundreds of litres of water. Use food-grade HDPE plastic barrels. The barrels should be disinfected and cleaned with a dilute bleach solution, and then thoroughly rinsed before use – even if they were clean to begin with. Again, they should be stored away from direct sunlight in cool locations. If possible, empty, clean & disinfect and refill every 6 months. The water should still be safe to drink, even if it has been stored for several years, but it may well taste slightly of plastic from the storage container.

If the only water available to drink is contaminated by radioactivity, then it may be necessary to try to purify the water sufficiently to make it safer (or at least, less dangerous) to drink. How to do this, and how effective this will be, depends on the nature and extent of the contamination. Boiling the water will NOT purify radioactive contaminated water. Likewise, chlorine water purification tablets will not help (but may be useful for making natural, untreated water that is not contaminated, safe to drink.

The main options available for purifying radioactive water are distillation, ion-exchange and filtration through an activated carbon filter.

Distillation will remove solid contaminants, dissolved salts and radioactive contaminants such as uranium compounds but will be much less effective against volatile contaminants such as radioactive iodine and tritium. Using a radiation meter to monitor the water before and after distillation will reveal how effective or otherwise, the process has been in purifying the water.

Ion-exchange technology is an effective method for removing many decontaminants from water, but you must bear in mind that it is achieved by absorption of contaminants into the ion exchange column, and when the ion-exchange column is left unused for any period of time, these absorbed contaminants will leach back into the water reservoir of the column, and as their concentration will increase the more it is used, the concentration of contaminants in the stationary column of water when not in use will increase substantially above the levels in the incoming water. For this reason, it is important to flush the ion-exchange column before drawing off drinking water. Typically, you should disregard the first few columns' worth of water whenever the ion exchange column is used, before collecting water to drink. (So, if the ion exchange column holds 2litres of water, whenever you use it to purify water, discard the first 6litres of water that comes out of the column, as this may contain elevated levels of leached contaminants absorbed in previous runs).

Reverse osmosis is also deemed a very effective method for purifying contaminated water. Basically, a pressure difference is created across a membrane, which allows water molecules to be forced through the membrane, whilst preventing larger molecules from passing through. Contaminated water is placed on one side of the membrane, and clean water is drawn off from the other side. However, reverse osmosis cannot prevent small molecules such as dissolved iodine gas from passing through – so some potential radioactive contaminants may not be removed.

Water contamination has been a significant problem in the locality of the Fukushima nuclear power plant in Japan, since it was damaged, and authorities recommend that water is purified using a combination of ion-exchange, reverse osmosis and activated charcoal, for the most effective removal of radioactive contaminants.

18 OTHER ASPECTS OF NUCLEAR SURVIVAL

Shelter

After one or more nuclear detonations, it will be essential to shield yourself from radioactive fallout. Fallout will start to reach the ground around 10 -15minutes after the explosion and will be mostly deposited after 24 hours. The fallout is extremely radioactive, however the radionuclides with the highest activities are generally those with the shortest half-lives, so although the radioactivity will be extremely high in the short term, it will drop off fairly rapidly. You MUST stay indoors or in a nuclear shelter for at least the first 24 hours, and it is highly advisable to remain there for a period of days – maybe up to two weeks – but listen to radio, TV or internet broadcasts from authorities for advice. Radiation levels will be sufficient to cause serious harm for a long period of time – maybe for up to five years, but after two weeks, levels will probably have dropped low enough for you to evacuate to a safe location. Once you have the all-clear to do so, this should be done as quickly as possible to minimise your exposure to radiation during travel, and the authorities will hopefully organise mass evacuation of affected populations.

If civilians start to self-evacuate, there is a high chance that the roads will be gridlocked, and there is no sense in sitting stationary in a grid-locked vehicle for hours or days, causing unnecessary exposure to radiation and potentially also food/drink shortages – it may be better to wait an extra 24 hours inside your shelter, by which time the exterior radiation levels will have dropped to a lower level, and hopefully traffic cleared, reducing your journey time and therefore your cumulative radiation dose.

The ideal radiation shelter will be a few metres underground – this will provide you with radiation shielding in every direction. It will also provide excellent blast resistance, which will prove invaluable if you are forewarned about the nuclear attack and have time to get inside your shelter. There are many designs for underground shelters, and some companies even manufacture fully assembled shelters that simply need to be buried in a pre-dug hole. Such shelters are often based on horizontally-aligned cylindrical designs – providing strength and utilising off-the-shelf plastic, concrete or galvanised cylinders, often based on manufactured components originally intended for water pipelines or septic tanks. To provide a suitably sized, flat floor, a false floor needs to be installed above the lowest point of the cylinder. With preplanning, this can be utilised as an under-floor storage space for food, water and other supplies, accessed through various hatches in the floor. Such underground bunkers would need to be planned to ensure adequate

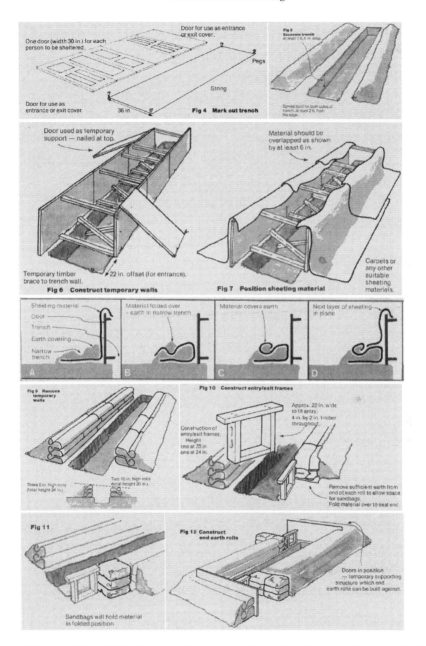

Figure 21: Initial stages of construction of an improvised radiation shelter, taken from a UK government publication 'Domestic Nuclear Shelters'. Reproduced with kind permission of the National Archives Office. (Contains public sector information licensed under the Open Government Licence).

ventilation, adequate filtering of ventilated air, sanitation, sewage, washing facilities, cooking facilities, electrical power storage etc. The construction of a decent underground bunker, complete with all the desired amenities, requires design and preplanning, as well as a fair amount of work, commitment and expenditure. It would need to be completed in advance of a radiation emergency and would realistically take a minimum of several weeks to construct and kit out - not something that you can easily create rapidly, or safely during a nuclear emergency.

If you have 1-2 days prior warning, perhaps because there has already been a nuclear strike but not in your locality, there is time to build an improvised shelter. Various designs have been produced by the American and UK governments for quick, make-shift shelters using available material, and are available online for free. Anyone can build such a shelter, either sunk into the ground or, if the ground is water-logged, it can be constructed above ground, banking soil up the sides and over the top. In the absence of specific building materials, the sides and roof of the shelter can be constructed from interior doors removed from a property. Carpet can be taken up from inside a building, and used to form a barrier against which earth can be built up – over the roof and against the walls. Construction notes for such a building can be seen in Figures 21 and 22.

For many people, it makes more sense to use what you already have available, and if you are not prepared, you will have no option other than to use whatever is available to you at the time. If your house is still standing, retreat to the basement if you have one, or if not, to a room in the interior furthest away from exterior walls and from the roof. Make

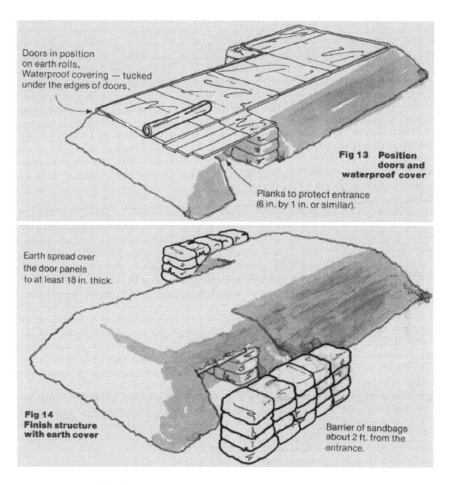

Figure 22: Final stages of construction of *an improvised radiation shelter, taken from a UK government publication 'Domestic Nuclear Shelters'. Reproduced with kind permission of the National Archives Office. (Contains public sector information licensed under the Open Government Licence).*

efforts to seal off the property, as described in the *Short Term Actions* section of Chapter 8, to reduce the amount of fallout that can enter the property.

If your house has been destroyed by a nuclear explosion, or has been made too unsafe to inhabit, your best option will be to make your way, rapidly, to the nearest standing property that has survived the blast. If there are no surviving structures that are capable of being reached within the 10-15minute time frame that may be dictated by fallout, then your only option will be to adapt something, but you would have very little time to do so. Before fallout reaches the ground, and in the initial stages of fallout, your main enemy is the high levels of radiation – so any natural feature may provide some protection – such as pitching some kind of sheet or tent up against a steep bank or cliff, or against a retaining wall or against the wall remnants of a collapsed house. Whilst sheltering, you can dig a trench and take refuge in the trench itself, to take advantage of the extra side-shielding provided by the walls of the trench. The deeper the trench the better, but even a shallow trench, that you can lie or crouch in, which could be excavated in 30 minutes, would provide some minimal protection, especially if you can cover it with something more solid– e.g., timber, or roofing sheets. If you can throw the excavated soil on top of this makeshift roof, you will provide valuable additional shielding. Even when located within a shallow trench, you can continue to excavate whilst sheltering at the same time. Of course, without a filtered ventilation system, you will still be in danger of breathing fallout, and fallout may enter the trench in the air, but even a small amount of protection over your head will significantly reduce the amount that enters. If you have a P3 radiation mask, then you should wear it. If not, any kind of mask will provide an element of protection, or even using cloth or clothing to shield your nose and mouth will have some effect in reducing inhaled contaminants.

Security

In the aftermath of a nuclear war, there is a predictable loss of social structures, government control, law and order. Disasters on a smaller scale have demonstrated that chaos and lawlessness quickly descend, with associated theft, violence and looting. You need to ensure that your supplies are safely stored and ideally hidden, to avoid someone else taking advantage of your advance planning.

In desperation, even citizens who would normally be law-abiding, might be tempted to resort to crime if it meant the difference between eating and not eating. Shops, supermarkets and pharmacies would be looted very early on. After this, desperate people will turn against their neighbours, or someone else's neighbours as a potential source of basic necessities. It goes without saying that anyone who appears to be well-provided for is likely to attract interest and become a target for looters. It is therefore important to secure any belongings that you wish to retain. Security may be afforded in three ways – 1) Hiding intrinsically valuable goods, and anything that is valuable to your survival – food, medicine, drinking water, tools etc., so that it is not found by anyone looking to loot your property, 2) Protecting your valuables, and your shelter, with security and by force if necessary, 3) By not giving the appearance of someone who has adequate provisions – you will need to look as desperate, and unprovisioned as everyone else.

Weapons

Many survival texts rather dwell on the subject of weapons. Without a doubt, weapons will help secure your property and belongings, but they will also suggest that there is something worth stealing. Firearms are clearly a powerful deterrent, and under the US Constitution, there is a right to bear arms, meaning that anyone can buy a firearm – with a few exceptions – namely users of illegal drugs, convicted criminals who have been handed sentences of over 12 months duration, or those who are accused of a crime punishable by a prison sentence of over 12 months, anyone dishonourably discharged from military service, anyone involuntarily committed to a mental institution, or convicted of domestic violence, or having a court order against them in respect of stalking, harassment or threatening behaviour. In the UK firearms are much more tightly controlled, and applicants need to demonstrate a requirement to possess a firearm, as well as not having a criminal record or any record of mental illness, depression or domestic violence. Applicants are carefully vetted both in terms of their medical history, criminal history and with character references. For that reason, legally owned firearms are much less common than in the USA. Therefore, in the UK, non-licensed legally owned weapons are limited to low-powered airguns, bows and arrows, crossbows, knives, catapults etc. It must be stressed that it is illegal to use any weapon offensively or to carry a weapon as a means of self-defense in the UK.

All weapons should be properly secured to prevent them from being stolen, and to prevent them from being accidentally discharged or used by children. Firearms should be concealed in a purpose-built, secure, lockable cabinet, ideally housed somewhere where it will not be obvious to an intruder. The keys should also be kept secure, ideally in a combination key-safe, and the combination code kept secret Also, bear in mind, that your weapons can also be used against you by an attacker, who may be much less wary of causing you serious injury, than you are of causing them serious injury.

Communication

Communication is essential in the modern world. Youngsters spend hours every day using social media and the internet. In the event of a serious local emergency, then continuity of online communications is likely to be unreliable. In the event of a nuclear war, it is likely that internet, mobile network and landline networks may well be compromised or non-existent, and the power network that supplies power to the transmitting stations, cellular towers and telephone exchanges may have been destroyed. These places may have back-up batteries, which will provide power for one or two days at best. Telephone exchanges and TV / Radio broadcasting stations may have a back-up generator, but this will only provide power for a few days, by which time the generators will need refueling – which will not be possible if the road network is badly damaged or if the fuel supplies are disrupted and people are panic-buying whatever fuel they can get their hands on – all are very likely scenarios. In such circumstances, you may suddenly feel cut off from the outside world. You may not be able to telephone to check if other members of your family are ok. In a society that relies on centralised communication systems – telephone exchanges, cellular networks etc., these become very obvious targets for an enemy that wants to create confusion and disarray. In these circumstances, there is a good solution – personal radio transceivers. These do not require any infrastructure, so unless an enemy can destroy each and every small radio owned by civilians – an impossible task – this system of communication will be highly reliable in the case of an emergency. Licence-free handheld radios can be purchased very cheaply in most countries, although rules and band allocation will vary from one country to another. Modern handheld radios are normally low power (e.g., 0.5W), frequency-modulated ('FM'), and utilise high-frequency bands such as the 446MHz frequency band. Battery life will potentially exceed 10 hours either using single-use AA or AAA cells. You can use rechargeable NiMH cells with many brands, and some have built-in NiMH or lithium ion battery packs. Sound quality is reasonably good, and the range would be as high as 15miles if you have good line-of-sight to the receiving location, or as low as 2miles in built-up areas such as towns and cities. Handheld radios such as these are great for keeping in touch with friends and family locally, but not useful for longer distances.

If you want to use more powerful, longer distance communications, then you can consider the CB ('Citizens Band') radio, which operates in the 27MHz band (although the USA has slightly different frequencies allocated compared to the UK, rendering equipment incompatible between the two countries). CB radios are limited to 4W maximum power output in the USA, UK and Europe (or 12W for SSB (single side band) modulation). As a general rule, lower frequency radio waves used by CB radio, can travel further, and are able to pass through buildings and trees more easily, but are not able to carry as much information (or sound quality) compared to VHF and UHF bands ('Very High Frequency' and 'Ultra High Frequency'). No licence is required to operate CB radio in the USA or the UK, but you will still need to comply with various regulations, particularly to ensure you do not create radio interference. Handheld CB radios generally have a very poor range – maybe up to a mile, but vehicle-installed radios, which often use a large whip antenna, often magnetically attached to the roof of the vehicle, can achieve ranges up to about 5miles. So called 'base stations' operated from a building with a sizeable antenna mounted high up, can achieve a range of 10-30miles depending on the location and topography. Some CB enthusiasts use amplifiers (often known as 'burners' or 'boosters') to increase their power output considerably over the legal limit, and thus increase their range, maybe up to 50-60miles, but very much dependent on having a decent antenna setup and a good, high location. Although CB amplifiers are readily available, and not illegal to own, it goes without saying that operating at power levels greater than the 4W limit is illegal, and potentially you could be fined, have your equipment confiscated or even go to prison. Having said all of that, the authorities tend to have better things to do with their time, so prosecutions are few and far between – generally limited to operators who are causing a nuisance or interference.

For those who wish to have an improved communication range, even when using low power operation, the obvious choice is to sign up as a Radio Amateur (also known as a Radio Ham). Radio Amateur bands are assigned in almost all countries at a wide range of frequencies – from very low frequencies (135kHz) right through to extremely high frequencies (over 275GHz). Radio Amateurs exist all over the world, and most countries assign the same or similar frequencies to the commonly used bands. You normally need to pass a technical exam to obtain a licence, and there may be a small payment required in order to sit the exam. Once you have passed the exam, it is free to apply for and obtain a licence in the UK, but in the USA, from 2022 there has been a fee for obtaining a Radio Amateur licence, although it is not excessive. Most countries now operate various classes of Radio Amateur licence, so do not be put off by the prospect of having to sit a technical exam – the lowest classification of licence (which allows you to operate up to 10W in the UK or considerably more than that in the USA) is not very technical at all and can be learnt fairly easily in a short period of time. Various books are available which will teach you what you need to know to pass your Radio Amateur exam without too much trouble. Further information is available from the ARRL in the USA or the RSGB in the UK.

Becoming a Radio Amateur gives you greater opportunity to communicate, both on a local scale and reliably over longer distances. Many transceivers allow you to choose between multiple radio bands – so if conditions are poor on one band, you have several alternative options. Even running a ham radio at 10W – the same power as a domestic LED lightbulb, Radio Amateurs regularly communicate over distances of several hundred miles by bouncing their radio signal off ionised layers of air in the atmosphere, and, under ideal conditions, can achieve communications of several thousand miles. Amateur Radio bands are well-used, and it is easy to locate other Amateur Radio enthusiasts in your area, or in other countries around the world, so in an emergency, you will be able to gather information worldwide. Many countries allow unlicensed operators to use Amateur Radio for emergency purposes.

One of the fundamental principles behind Amateur Radio is to assist during emergencies - to provide communications help, and to relay information for emergency services during disasters. Although you may think that this is unlikely to happen in this day and age, it happens regularly across the world, and there are several recent examples in the USA – for example, after the Boston Marathon bombing of 2013, Radio Amateurs provided communications when the cell phone network became overloaded. Similarly, Radio Amateurs provided valuable communications services in 2005, when Hurricane Katrina left over three million people without basic utilities, some for an extended period, and in the aftermath of the 9/11 terrorist attacks on the World Trade Centers in New York.

Energy (Fuel for Heating, Vehicle, Generator)

Depending on the time of year, fuel for heating may be essential for survival, or not required at all. Houses whose only form of heating relies on an electricity supply are particularly vulnerable – storage heaters and heat pumps, for example, only work when there is a mains electricity supply, but central heating systems based on gas or oil fuels also require electricity - to run the boiler electronics, and circulation pumps.

Stoves that run on coal or wood provide much greater independence. Of course, if the stove has a back-boiler that is pumped rather than relying on 'gravity circulation', then it is still electricity dependent. If at all possible, it is a good idea to have a solid fuel room-heating stove. If the stove has a back-boiler to heat a hot water cylinder using gravity circulation, then it will provide hot water and some heating regardless of whether there is an electricity supply. However, it should be cautioned that if the water supply fails, then it will no longer be able to provide hot water.

Figure 23: Various radio transceivers: At the bottom of the stack is an early 1980s 4W CB radio, on top of which are stacked two multiband amateur radios - the G90 is a 20W radio transmitting on bands from 1.8MHz to 30MHz, and the G1M on top of it has an output of 5W from 3.5MHz to 30MHz). The silver handheld radio transmits 0.5W, licence-free on 446MHz. The Baofeng amateur radio on the right transmits 5W on two bands at 144MHz, and 440MHz.

If the nuclear incident is close enough that there is a danger of radioactive fallout or radioactive gases in the air – which you can ascertain from public broadcasts and through using your radiation monitor, then you should not run solid fuel, liquid fuel or gas stoves as it will increase the speed at which air-changes occur within the house, unless they are of a design where air is drawn directly from the exterior of the property, and the exhaust is vented to the exterior of the property (for example, 'balanced flue' boilers). Alternatively, if you are cold, use additional clothing or blankets to keep warm.

Stockpiling liquid (or gaseous) fuel for vehicles or for running generators to provide electricity will give you a big advantage if fuel restrictions occur – which is quite likely in any kind of national emergency, but there are rules and regulations about the quantity of fuel that can be stored, and the manner in which it is stored (and precautions that need to be taken to prevent leakage into the environment), which can restrict the amount you are allowed to store, or make it difficult and expensive to store a larger quantity.

Heat for cooking and hot drinks.

The need for heat will eat into your fuel stores, so anything you can do to reduce your fuel dependence will help conserve resources. If it is safe to go outside, then food can be cooked in solar ovens – these are easy to fabricate yourself, from a cardboard box, lined with silver foil. The flaps of the box – also lined with foil, are positioned to reflect additional sunlight into the box. The food is positioned at the focus point, where it will heat up and gradually cook. There are variations with and without glazing, and the ovens can reach 150°C on a cloudless day. However, you will need to regularly reposition the cooker to face the sun during cooking and on cloudy days, the temperature is substantially lower.

Solar Vacuum Tubes.

Solar heating can be achieved in many ways. Greenhouses are a good, simple example of solar heating – glass is opaque to shorter wavelengths of infra-red radiation, and if the sun is shining, the interior of the greenhouse will increase in temperature, and on a bright day, will be substantially warmer than outside. This phenomenon is often referred to as the 'greenhouse effect'. This effect is utilized in most designs of solar 'collector', where glazing at the front of the collector traps the solar infra-red radiation, which is absorbed usually by a dark painted metal behind the glazing. However, as the absorber gets hot, it heats the air inside the collector, which in turn heats the glass, which then radiates heat away. The bigger the temperature difference between the exterior and the interior of the solar collector, the lower its efficiency becomes.

One ingenious solution to this is to remove the air between the glazing and the absorber. However, due to the pressure of the atmosphere, the force exerted over a few square metres of solar collector is enormous and would result in implosion of the glass. However, if the surface is small and curved, it is much better able to cope with the pressure difference. This is achieved in a vacuum tube solar collector- imagine a thermos flask, stripped of its outer casing, and the inner wall coated in a material that is very efficient at absorbing heat, and very inefficient at emitting heat. If you place water inside the flask, it can be heated by the sun, without the need for pumps, controllers, or any moving parts.

In 2007, I demonstrated a novel method of cooking, using vertical vacuum solar tubes, in various configurations, to achieve higher temperatures, especially in overcast conditions or when ambient temperature was low. I demonstrated, with photographs and video, cooking curry in a single solar tube, which was posted on a popular forum. This prompted a discussion in which various participants suggested improvements or variations to the idea. The vacuum tube cooking method proved reliable even in overcast conditions, and a single tube can heat almost 2 litres of water to boiling temperature on a bright day, or to 70+°C on an overcast day. For cooking purposes, the vacuum tube can exceed 120°C in bright weather. It is also possible to construct an insulated oven connected to several tubes, which can provide a dry heat for baking.

Figures 24 & 25: Solar vacuum tube reaching 115°C on an overcast day, in the UK.

Figure 26: Solar tubes used as kettles for serving tea and coffee at a show in 2007. Even on heavily overcast days, we could serve hot soup to willing volunteers!

There are now many vacuum tube solar cookers available on the market, usually incorporating much shorter tubes for easier handling.

Disadvantages

- The obvious disadvantage of any form of solar heating is that it is dependent on the weather – bright days are better than dull days!
- The solar tubes are made of glass, and therefore can easily be damaged or broken if handled carelessly, or if blown over by the wind.
- At the base of each tube is a weak point, where the tube is sealed after the vacuum is formed during manufacture. It is very easy to damage this seal, and therefore tubes should ideally be capped at the base with a rigid plastic cup to protect the seal and prevent damage.
- The tubes can also be shattered by thermal shock – for example by pouring cold water into an empty tube that has heated by virtue of being left exposed to bright daylight.
- The tubes can be damaged by excessive mechanical vibration (for example, if transported by vehicle over rough ground in insecure packaging).

However, if care is taken, vacuum solar tubes can last indefinitely. Care should always be taken to ensure that the tube is not tightly capped, otherwise tube damage may result from expansion or boiling of the contents.

Figure 27: An 'integrated solar collector' – direct heating solar vacuum tubes heat water which rises through convection into a large, insulated tank at the top. This stores the hot water until it is required.

In addition to provision of heat for cooking or for hot drinks, solar tubes can be used to provide hot water for washing. In an emergency, a single tube can be employed to heat small amounts of hot water for washing. Alternatively, solar heaters exist which use multiple tubes connected to an insulated header tank which acts as a hot water store – see *Figure 26*. These are very commonly used in countries such as Greece, China and Turkey although in Europe and the USA, more compact designs connected to internal heat stores are generally used.

Very simple water heaters exist which consist of a black plastic container, connected to a hose with a shower head attached to the end. These are not particularly efficient, but work adequately on a sunny day, and are certainly better than showering in cold water! It would be very easy to construct such a heater in an emergency situation by painting any small container black and leaving it in the sun to heat.

Source of Electricity

Generators can provide electricity to run most household items. Generally, they fall into two classes:

1) Spark ignition engines – which can run on or be adapted to run on petrol, methanol, ethanol, butane, LPG (propane), Natural Gas (methane) or hydrogen.

2) Compression ignition engines – are generally run on 'road diesel', 'agricultural diesel', kerosene, biodiesel or can be modified to run on vegetable oil. There are some engines (and indeed vehicles) which will run on undiluted vegetable oil ('SVO' – straight vegetable oil). Usually, the only problematic issue is starting the engine – you either need to start on road diesel, and then switch over the vegetable oil when the engine is hot, or you heat the vegetable oil before it is injected. Most modern engines with common-rail injectors will not run on vegetable oil. The chances are that vegetable oil will become unavailable just as quickly as diesel does, in the event of a disaster.

Unless you have a particularly large generator, you will need to take care not to run too many energy-hungry electrical appliances simultaneously.

Generators running continuously will get through large amounts of fuel. For example, a diesel 5kW generator will use 2-3 litres of fuel per hour and a petrol 5kW generator will use 3-5 litres of fuel per hour - partly due to the lower energy value per litre of petrol vs diesel, and partly due to the slightly lower efficiency of the petrol engine. In fact, under full load, the petrol engine efficiency will be pretty similar to that of the diesel engine, whereas at part load, the diesel generator will be significantly more efficient. All things considered, if you plan to run a generator for say 5 hours per day, you will need to store over 100litres of fuel for each week you plan to rely on your generator. So, if you're planning to rely on a generator for up to 3 months, you may need to store over 1200 litres of fuel. This is potentially hazardous, may not be legal (depending on the fuel and your location), and is expensive to do safely. Fuel is likely to be considered a valuable resource in the event of an emergency and would be a prime theft target.

Photovoltaics and batteries

The war in Ukraine has demonstrated how attackers may deliberately target civilian power as a strategic objective, often repeatedly.

In a nuclear war, the power distribution network and the power stations themselves may be damaged or destroyed by the force of the nuclear blast, even without a targeted attack. Add in the probability that the environment may be too contaminated for people to go to work at the power stations, and the likelihood that fuel supply chains will be interrupted or destroyed, and the outcome is almost certainly going to be rolling blackouts, large areas permanently cut off from electrical power or no power whatsoever.

Therefore, it makes a lot of sense to have some form of renewable power back-up, based on solar, wind or water. The downside of course is that these will all be potentially adversely affected by EMP. Conventional wisdom dictates that anything connected to long wires will be particularly susceptible, so PV that is wired up to an inverter/battery or inverter/mains setup will experience issues. Conventional wisdom also dictates that the panels are likely survive EMP if they're not connected to anything 'because there are no electronics' – however all larger PV modules will have bypass diodes in the junction box on the back of the panel – usually three diodes, and less commonly two or four diodes. The purpose of the bypass diode is to isolate under-performing sections of the PV module – for example if part of the panel is producing significantly less power than the rest of the panel due to shading from a chimney, telegraph pole or tree branch. The incorporation of bypass diodes increases the overall efficiency of a system in situations where transient shading occurs. Very small PV panels (50W or below) generally do not have bypass diodes. Tests carried out suggest that there is a possibility that the bypass diodes could fail during EMP.

Some military research has been carried out on this, concluding that bypass diodes are indeed susceptible, particularly to reverse surge current, but that the PV cell junctions are relatively immune. It is suggested that the problem can be overcome by installing Transient Voltage Suppressor (TVS) diodes, to protect each panel ideally, or if not, each circuit ('string') of panels. Alternatively, it is entirely feasible to simply remove or replace the bypass diodes in the panel junction boxes after an EMP event. The associated electronics, such as the PV inverter can be fairly easily protected by shielding the PV feed cables (e.g., by installing these inside earthed metal trunking, and the associated electronics can be shielded by installing within a metal cabinet. The incoming supply cables would also need to be protected, assuming the system was connected to the grid at the time of the EMP, and this is achievable using HEMP (High Altitude EMP) power line filters – which are already available as a manufactured item.

Many households have PV systems these days, normally feeding power directly into the 240V or 110V AC electric supply. In the event of an EMP and the elimination of the power grid, it would be necessary to run the PV as a stand-alone system into batteries. With the advent of lithium-ion household battery banks, this is certainly feasible, but most of the systems on the market do not allow for battery charging during power outage. So, the system would need updating to allow this, or reconfiguring as a stand-alone battery system – something that is not particularly straightforward with high-capacity lithium ion battery systems. Alternatively, a small-scale stand-alone PV system could be employed, separate to the main household PV system, for use in emergencies. This could be as simple as a single solar panel connected to a USB-converter or 12v regulator, to allow small low-power battery-based appliances to be charged-up as necessary. Lead acid batteries have several disadvantages relative to lithium-ion such as higher self-discharge rates, lower energy storage capacity and heavier weight. However, they are simple and rugged technology, and are better able to cope with poor charging regimes and overcharging. In an emergency, you could easily remove the battery from your vehicle and use this with a small PV panel to provide approximately 1kWh of electricity storage, for running radios, LED lighting, and for recharging mobile phones and flashlights, requiring only the addition of a 12v-to-USB converter, which are readily available and cheap.

Figure 28: A large roof-mounted PV system using monocrystalline solar modules in the UK.

When No One's Coming

19 CONCLUSIONS

Surviving a nuclear disaster is a daunting task, but it is possible with the right knowledge and preparation. There are several threats and hazards associated with nuclear disasters that one must be aware of in order to stay safe. These threats include radiation exposure, fallout, and the destruction of infrastructure and communication networks. In this book, we have explored these threats in detail, and provided practical advice on how to mitigate their effects.

One of the most significant threats associated with nuclear disasters is radiation exposure. Radiation can cause a range of health problems, from skin burns and radiation sickness to cancer and genetic mutations. To protect oneself from radiation exposure, it is important to understand the different types of radiation and their effects. For example, gamma radiation is the most penetrating type of radiation and can travel long distances, while alpha radiation can only travel a few centimetres and can be stopped by a piece of paper. Understanding these differences can help you take appropriate protective measures, such as using shielding materials like lead or concrete.

Another threat associated with nuclear disasters is fallout. Fallout is the radioactive material that is released into the atmosphere during a nuclear explosion or accident. Fallout can travel long distances and contaminate the environment, including food and water sources. To protect oneself from fallout, it is important to shelter in place or evacuate to a safe location as soon as possible. It is also important to have a supply of clean water and non-perishable food on hand in case you need to stay in your shelter for an extended period of time.

In addition to these threats, nuclear disasters can also cause significant damage to infrastructure and communication networks. This can make it difficult to access basic necessities like food, water, and medical supplies. It is therefore important to have a plan in place for evacuation or sheltering in place, and to have a supply of essential items like food, water, and medical supplies on hand.

To survive a nuclear disaster, it is also important to have a basic understanding of survival skills. These skills include first aid, water purification, fire building, and navigation. Knowing how to treat common injuries, find and purify water sources, and start a fire can mean the difference between life and death in a survival situation. It is also important to be able to improvise and adapt to changing conditions in order to stay alive.

In the aftermath of a nuclear disaster, it is important to follow the guidance of local authorities and emergency responders. This may include staying indoors, avoiding contaminated areas, and taking steps to decontaminate oneself and one's belongings. It is also important to have a plan in place for communication with family members and loved ones, as communication networks may be disrupted in the aftermath of a disaster.

Ultimately, surviving a nuclear disaster requires a combination of preparation, knowledge, and resilience. By taking steps to protect oneself and one's family, and by staying calm and focused in the face of adversity, it is possible to survive even the most catastrophic of events. While it is impossible to predict when a nuclear disaster may occur, by taking the steps outlined in this book, you can be better prepared to survive and recover from such an event.

When No One's Coming

Figure 29: Gas masks awaiting Armageddon.

When No One's Coming.

APPENDIX I: EMERGENCY PLAN TEMPLATE

Emergency Preparation Plan

Who this plan protects:

Name	Age	Contact Details	Specific Needs / Medication	Social Security Number
1				
2				
3				
4				

5				
6				
7				

Pets (Number and type):

Home Address:

Extended family / Neighbours to help if possible:

Scenario	Anticipated Duration of Scenario
Extended power outage ☐	_____Days/Weeks/Months
Extended heating outage ☐	_____Days/Weeks/Months
Civil unrest / Riots / Looting ☐	_____Days/Weeks/Months
Personal Emergency / Serious Injury ☐	_____Days/Weeks/Months
Economic instability / Job loss ☐	_____Days/Weeks/Months
Flooding ☐	_____Days/Weeks/Months
Heavy Snow ☐	_____Days/Weeks/Months
Hurricane ☐	_____Days/Weeks/Months
Tornado ☐	_____Days/Weeks/Months
Wildfire ☐	_____Days/Weeks/Months
Earthquake ☐	_____Days/Weeks/Months
Pandemic ☐	_____Days/Weeks/Months
Nuclear Powerplant incident ☐	_____Days/Weeks/Months
Conventional War ☐	_____Days/Weeks/Months
Nuclear War ☐	_____Days/Weeks/Months

Emergency Assembly Points or Shelter Options:	
Indoors:	
Outdoors:	
Alternative accommodation & Distance:	
Alternative Location: (If none of above are possible)	

Evacuation Plan:

Location of Escape Bags:

Evacuation Options:	
☐ On foot ☐ Bicycle ☐ e-bike ☐ e-scooter ☐ Other:	☐ Motorcycle ☐ Car ☐ Off-road Vehicle ☐ Other (_____)

Evacuation Routes:

Communication Plan:

Shelter in Place Plan:

Preparation Checklist:

Prepared Escape bag for each person? ☐
All family members aware of this plan? ☐
Have you planned how to communicate with each other if separated? ☐
Have you prepared emergency medical supplies? ☐
Have you prepared emergency food and water supplies for the duration anticipated? ☐
Have you prepared emergency supplies of medication for the duration anticipated? ☐
Does every person have a facemask & suitable filter? ☐
Note important contact details and addresses ☐
Have you made adequate provision for pets? ☐
Have you planned a location for a food / water store? ☐
Do you know how to build a shelter ☐
Is at least one person trained for first aid? ☐

Checklist if you have to leave your home:

Wear comfortable, sturdy clothing and shoes suitable for hiking and take waterproofs and extra layers ☐
Pack a raincoat and protective gloves, and a mask ☐
Turn off Utilities ☐
Hide valuables & ensure that other people in your family know where ☐
Hide any emergency stockpiles that you cannot take with you ☐
Lock all windows and doors. ☐

List of tools you may need (list by category):

Emergency Contacts:

Food Stockpile Inventory

Food	Storage Conditions	Manufacturer Use-by Date	Revised Use-by Date	Date for Rotation

Date of Next Review / Revision: _____

APPENDIX II: THE NUCLEAR EMERGENCY CANISTA

The Nuclear Emergency Canista™, available at *www.10secondstomidnight.com* is a grab-pack containing selected PPE, radiation monitoring equipment, medical supplies, iodine supplements, a solar panel and hand-crank for charging devices, and survival supplies - all in a large, waterproof, sealed, easily transportable container, for use in a nuclear emergency. It is not intended to be an all-encompassing inventory, instead a pack containing a selection of useful tools, equipment and consumables, all in one place, ready to access at a moment's notice. The contents of a typical Canista™ is listed below, with suggested uses or purpose for each item. Customers can vary the contents of Canistas to their own specifications.

Radiation Monitor (Geiger Counter)	A dual purpose sensitive digital radiation monitor and dosimeter that displays ambient radiation levels as well as a cumulative dose. Used for establishing radioactive safety of environments, identifying sources of contamination, for monitoring decontamination procedures, or for providing dosimeter readings to establish extent of exposure.
Full-face Respirator ('3M'* Bayonet Pattern)	Protective face mask for use in hazardous environments. The mask will accept '3M'* bayonet style filters.
2x Filters for Respirator (NATO 40mm)	P3-type Radiation filters for removal of radioactive particulates in air. If you are desperate for drinking water, you can use one filter canister as a water-filter to remove radioactivity from the water. Check water with Radiation Monitor before and after to ensure it is effective.
2 x Adaptors NATO 40mm to '3M'* Bayonet Pattern	These allow the 40mm NATO cartridges to be used with the 3M flanges on the full-face respirator.
Hand Crank/Solar-Powered Radio/Torch/USB Power supply	Hand-cranked dynamo provides a small amount of power which runs the built-in radio for monitoring news bulletins and emergency information transmissions. It also has an in-built torch to provide a dependable light

	source, and a USB output for charging mobile devices.
Disposable Hazmat Coveralls	For protecting body and clothing from contamination with fallout if you have no option other than to enter a fallout-contaminated environment, or if you need to decontaminate someone.
10W 12v Monocrystalline PV Panel	Perpetual power supply for running/charging low power 12v appliances. Can also be used with DC converter to produce 5V for charging/running USB devices.
12v to 5V USB HF converter for PV	For converting power from 12v PV panel for USB devices.
First Aid Kit	Emergency Foil Blanket, disposable Gloves, Tweezers, Medical Scissors, Instant Cold Packs, Sterile Eye Wash, Bandages, Triangle Bandages, First Aid Tape, Wound Dressings, Crepe Bandage, Sterile Eye pads, Safety Pins, Assorted Adhesive Plasters Non-Adherent Pads, Knuckle & Finger Fabric Strips, Antiseptic Wipes
Precision Scales (0.01g - 200g)	Useful for measuring much smaller amounts of material than with a standard kitchen scales. This is primarily provided for weighing out the potassium iodide supplement, which can then be dissolved in a glass of water or juice.

100 Water Purification Tablets (Each tablet contains 8.5mg of Sodium dichloroisocyanurate)	One tablet will purify 1litre of water, if it is of unknown quality. ONLY SUITABLE FOR BIOLOGICALLY CONTAMINATED WATER!! These will not make radiologically contaminated water safe to drink. However, in the absence of safe drinking water, if you can locate water that has not been contaminated with fallout, these will provide you with access to water safe for drinking. Instructions for use: 1)Add one tablet to 1litre of water 2)Stir to distribute evenly 3)Wait 30minutes before drinking.
Disposable Gloves (100pack)	These gloves will protect your hands from biological and radiological contamination. Wear them when administering first aid to casualties, when you are forced to enter contaminated environments and when handling potentially contaminated materials/clothing etc. If you are planning on carrying out hazardous tasks, it is advisable to wear two pairs of disposable gloves, one over the top of the other, as they are not hard-wearing and can tear. If you are carrying out physical tasks, and you have work gloves available, wear these over the top, but bear in

	mind that they will get contaminated.
Heavy Duty Cling Film	This has all kinds of uses. It is very strong when wrapped around objects, has good electrical insulation, and is relatively impermeable to wind and weather. It can be used for: 1) Wrapping electrical items before wrapping in foil to protect against EMP (and over the top of foil to keep it in place and to prevent it ripping). 2) Sealing off air vents and fireplaces to reduce the amount of contaminated external air entering a house. 3) Sealing broken windows/doors to prevent contaminated air entering house. 4) Tying things down or together. 5) Makeshift rope (double-up as necessary to provide required strength). 6) Makeshift contamination-protection for boots and shoes (Will need several layers, but be mindful that it will wear off the soles quickly, especially on abrasive surfaces) 7) First Aid – substitute for triangle bandages. 8) First Aid – for attaching splints for broken bones. 9) First Aid – for applying to burns.
Wide Sticky Tape	Useful in conjunction with the heavy-

	duty cling film for sealing-off air vents, chimneys etc. Also useful for binding small items together.
Heavy duty Rubbish Bags	Can be used to store things or to keep items protected from damp and weather. Also, for immediately collecting and sealing contaminated clothes or contaminated items.
Cable Ties	Can be used to quickly seal bin bags, also for tying down and securing.
Disposable Hooded Protective Suits	Can be worn over the top of clothing when it is necessary to enter a contaminated environment. Use in conjunction with full facemask respirator and protective gloves. After fitting facemask, pull hood over head to help protect against contaminating hair, ears etc.
Candles	Tealight candles are provided, as they do not require candleholders, and are less likely to fall over. Candles store exceptionally well, are inherently waterproof, and provide several hours of lighting.
~5 litres bottled water	Bottled water provided for short-term drinking requirements until alternative supply can be sourced. Will provide at least 24hours supply for 2 people.
Dark Chocolate Bars	These pack a lot of calories in a small size and have long shelf-life. Milk chocolate is also good and has a

	reasonable shelf-life but not as long as dark chocolate, due to the milk content. The best-before date on dark chocolate is usually around 2years, but it should be perfectly safe to eat it after 10years. You may see blooms on the surface of the chocolate if has been stored for a long time, but this is due to migration of some ingredients to the surface. If it smells ok, it is safe to taste it. If it tastes ok, it's safe to eat it. Gone-off chocolate tends to smell either mouldy, like sour milk or like garlic.
Potassium Iodide crystals (25g)	Provides emergency protection against radioactive iodine, which will otherwise be concentrated by the thyroid gland. Start taking potassium iodide as soon as possible after a nuclear incident occurs that puts you at risk, or 4hours before, if you have advance warning. Adults should take 130mg (0.13g) per day, so this will provide at least 48days protection for 4 people
Radsafe Liquid Soap	Radsafe is suitable for use as an aid to nuclear decontamination. It may also be used for general purpose washing of hands/body/hair. It is a misconception that a specific type of soap is needed for nuclear decontamination – any mild soap and

	water will do a good job of removing contaminants.
Extra Thick Aluminium Foil	This is used to protect electronics from EMP: First wrap the device in a couple of layers of insulating material (e.g., heavy duty cling film or paper), then wrap with 5 layers of the aluminium foil, ensuring that all parts are completely covered with each layer. When it is fully wrapped, either secure with tape, or with another layer of cling film to provide added protection to prevent the foil from tearing or snagging.
Lighters and matches	For starting fire / lighting candles / sterilising needles
Wet wipes (3 packs)	For general hand or body washing when water is not available. Also useful for decontamination procedures but be careful not to contaminate the whole packet.
AA and AAA Cells	Whilst rechargeable cells and a suitable charger are a better option for the long-term, rechargeable cells need to be regularly topped-up to prevent discharging, and if they are stored long-term without regular charging, the batteries tend to fail quite quickly. For this reason, single-use batteries are provided, which can be stored for several years. Alkaline batteries do NOT need to be

	protected from EMP.
Nuclear Survival Manual	The last thing you need to be doing during an emergency is hunting for your survival manual. Which is why there is one in the kit!
The Nuclear Canista(TM)	The actual container is waterproof, with a tight-fitting lid that seals. It can be used as a water store, or as a temporary toilet for use in bunkers/shelters (but not both!)

* '3M' is a registered trademark of 3M Company,

'Nuclear Canista' is a trademark of 10secondstomidnight

All trademarks, service marks, and company names are the property of their respective owners.

When No One's Coming

APPENDIX III: USEFUL PHONE APPS

There are various applications for mobile devices which may be very useful in times of emergency. This list identifies some which may be particularly useful in various emergency situations. It should be noted that whereas in some emergency situations, the cellular / mobile network maybe fully functioning and accessible, this will often not be the case – for example if the network has been taken down by NEMP, sabotage, conventional munitions, or power outage. If the network is still available, and particularly if it is operating at reduced capacity, it may be simultaneously overwhelmed due to the number of additional communications being attempted as populations panic and try to contact their loved ones, resulting in an intermittent, unreliable or non-existent service. You should therefore not rely on being able to access online services, so apps that can be used offline will be more reliable in the case of emergency.

1) Compass.

2) First Aid (e.g., Red Cross First Aid).

3) Chemical Hazard database (e.g., ERG2020 – hazard datasheets for most chemicals, with recommendations for emergency exposure treatment).

4) What3Words (alternative, simple location software based on a combination of three words to uniquely identify every $3m^2$ area of the planet – easier to remember and quicker to relay to others. It is also used by many emergency services).

5) GPS maps (offline mapping. Note that the GPS system may be disabled or destroyed by high altitude satellite busting NEMPs or by targeting with anti-satellite technology ('ASAT' or 'DA-ASAT').

6) Calculator.

7) FEMA Mobile App (USA).

8) Off-line Survival Manual App.

9) Panic button app (sends a message to designated contacts or to emergency services with your GPS position, in situations where you cannot speak or dial).

10) Life 360 tracker app (GPS monitoring of loved ones for quick location during an emergency).

APPENDIX IV: GLOSSARY OF TERMS

Absorbed dose: The amount of radiation energy absorbed by a unit mass of tissue.

Acute exposure: A radiation exposure event that takes place over a short period of time, as opposed to one that accumulates over a long period.

Acute Radiation Syndrome (ARS): The name given to the symptoms arising from short term exposure to a dose of at least 75rads of penetrating radiation. The first symptoms include nausea, fatigue, vomiting, and diarrhoea., but other symptoms such as hair loss, bleeding, swelling of the mouth and throat, and fatigue may follow. Exposure to over 1,000 rads may cause death within 2 – 4 weeks.

Air burst: A nuclear explosion at a high enough altitude that the fireball does not touch the ground. This is particularly relevant in terms of fallout, as the fireball will draw up earth, water, vegetation and other materials if it reaches the ground – which will be turned to radioactive dust, and is known as fallout, when it drifts back down to the ground. Because Air Bursts do not drag up material from the ground, they produce considerably less radioactive fallout.

Alpha particle: This is a charged particle made up of two neutrons and two protons, which is ejected at considerable velocity from the nucleus of some radioactive isotopes when they undergo nuclear decay. Alpha particles are one form of nuclear radiation. Alpha particles are not very penetrating and can be stopped by very small amounts of shielding, but they carry a lot of energy and can cause the most damage to human tissue.

Atom: The building blocks of all elements are atoms. An atom is the smallest amount of an element that can exist. These are the units that can be joined together to make molecules. It used to be thought that atoms were the most basic component of matter, but it was later discovered that atoms are constructed of even smaller components that are held together very strongly.

Background radiation: This is the radiation that is all around us, that we are exposed to constantly, every day. This radiation comes from various sources – the sun, other stars and galaxies, radioactive rocks beneath our feet and even a little due to residue from man-made nuclear explosions and accidents.

Beta particles: These are electrons which are ejected from the nuclei of some radioactive elements when they decay. Beta particles travel at high speed, although some are faster than others, depending on the isotope it is emitted from. Beta particles carry less energy than alpha particles but are a little more penetrating. Unlike alpha particles, Beta particles can penetrate exposed skin, and can cause radiation burns.

Biological half-life: For a given substance, it is the time taken for the concentration of a chemical that has been absorbed into the body through the skin, or inhaled or eaten, to be reduced by half, through excretion processes.

Catalysis: The process by which a material causes a chemical reaction to occur or to speed up - often by a very significant degree, but without being used up itself, by the reaction. Catalysts usually work by providing a chemical pathway which requires less activation energy. Sometimes catalysts work simply by binding two chemicals that may react with each other, so that they are in close proximity.

Chain reaction: In relation to nuclear processes, it is the name given to a nuclear reaction which self-perpetuates – so the products of the process (e.g., neutrons) are generated in sufficient quantity, to generate enough new events to keep the process going. A runaway Chain Reaction is identical, except in that each event causes more than one additional event, each of which in turn, produces more than one further event, thus the number of nuclear events increases exponentially. This is the basis of a nuclear bomb.

Chronic exposure: Is when a person is exposed to a substance over an extended period. Chronic exposure even to quite low levels of toxins can have a detrimental effect, if they are sustained over a long enough time period.

Contamination: The undesirable accumulation of toxic or radioactive materials on a person or object. The contamination may be external – such as being deposited on clothing or skin, or it may be internal, for example, by inhaling particles into the lungs, or ingested by eating contaminated food.

Convection: This is the upwards movement of a fluid, be it plasma, gas or liquid, due to the effect of heating. The reason is that, with very few exceptions, when fluids become warmer, they become less dense, and therefore experience less attractive force due to gravity than cooler fluid around it. As a fluid rises, it will often be cooled by, or mixed with, surrounding fluid, and as it does so, its rate of ascent will decrease, and it may then start to fall again. Convection can cause strong currents where large temperature differences occur – such as in the fireball created by a nuclear explosion.

Criticality: The state a fissile nuclear fuel is said to have reached, if it is able to sustain a nuclear reaction. This is where the number of neutrons produced by the reaction is more or less constant – with newly produced protons balancing any losses through absorption or ejection.

Critical Mass: The minimum amount of radioactive material that is needed to produce a self-sustained nuclear reaction.

Cumulative dose: The total amount of radiation that a person or object is exposed to over a period of exposure to radioactive materials.

Decay: Radioactive decay is when an unstable (radioactive) element undergoes a transformation into a different element or isotope, of lower mass. This results in the simultaneous emission of radiation.

Decontamination: The process of removing as much (radioactive) contamination from a person or object as possible, to reduce the associated hazard caused by the radioactivity.

Depleted uranium: This is natural uranium which has been processed to remove the more radioactive isotope uranium-235. The remaining uranium isotopes are less radioactive, and it is treated as non-radioactive by the military, who use it to create or tip projectiles. As uranium is very dense, it allows very heavy, compact projectiles to be manufactured, that have particularly good armor-piercing properties. There is, however, a significant health risk, and soldiers who are exposed to depleted uranium tend to suffer detrimental health effects in terms of chemical toxicity and mild radiological exposure.

Deuterium: This is an isotope of hydrogen in which the atomic nucleus contains a neutron, in addition to a proton. This form of hydrogen has double the density, hence its common name 'heavy hydrogen'. Deuterium is stable and naturally occurring. Together with Tritium, it is used as fuel for nuclear fusion reactors, and also in thermonuclear bombs.

Dirty bomb: A bomb designed to contaminate an area with radioactivity, but not actually undergoing a nuclear detonation. Dirty bombs use conventional explosives as a dispersal mechanism. It would be easier to make a dirty bomb than a nuclear bomb, and therefore concerns exist over the use of dirty bombs by terrorist organizations.

Dose: The amount of radiation that a person is exposed to.

Dosimeter: An instrument that measures how much radiation a person is exposed to in a given time frame. Dosimeters are often worn as monitoring equipment on outer clothing to record the cumulative dose of ionizing radiation that a nuclear worker receives in the course of his or her work shift.

Element: An element is a distinct material which cannot be broken down into anything else by chemical means. An element cannot be transformed into any other element except by the removal or addition of protons to its nuclei. For any given element, there may be several versions which have the same number of protons in each nucleus but may have different numbers of neutrons. These will all have the same chemical properties, but they will have very slightly different physical properties. Some isotopes are very stable and not radioactive, others may be unstable. Unstable isotopes will have a measurable half-life defining how long it takes them to decay. If an atom of a stable isotope is hit by a neutron, the neutron may be absorbed into the nucleus, causing it to become unstable, and in turn decay with the emission of radiation.

Enriched uranium: Uranium which has been subject to separation processes to remove some of the more mundane isotopes to leave a higher concentration of the naturally occurring uranium-235 isotope, which is more radioactive, and therefore more useful in nuclear reactors and nuclear bombs.

Fallout: Nuclear fallout consists of small particles of radioactive dust which start descending from the mushroom cloud of a nuclear explosion, reaching the ground 10-

15minutes after its initiation. Fallout will continue to descend for 24 hours, at which point the majority will have landed. Some fallout will be lifted high into the stratosphere, where it may take many months or years to descend to ground level.

FFP3: ('Filtering Face Piece 3'): This is a classification of breathing protection. This is the highest classification, designed to filter out 99% of particulate with a maximum leakage to the inside of 2%. Most FFP3 face masks will filter out particles down to about 0.3µm

Fissile Material: Nuclear fuel made up of heavy elements, whose nuclei are unstable, which may split into smaller nuclei, accompanied by the emission of nuclear radiation. Fissile material is used to produce atomic bombs, and for atomic energy production.

Fission: The decay of a large, unstable atom whereby the nucleus splits into two, separate, smaller nuclei plus an excess of energy which is released as heat and usually a few surplus neutrons. A percentage of these excess neutrons may collide and be absorbed by nearby nuclei, causing them to become unstable and subject to fission reactions.

Fusion: A reaction in which two low mass atoms join together to produce a single atom, emitting excess energy in the process. This is the process that the sun uses to produce energy and heat. It is also used in the later generation of nuclear bombs, also known as thermonuclear weapons.

Gamma rays: These are a form of high-energy nuclear radiation, emitted by some radioactive element when they decay. Gamma rays are very high energy electromagnetic waves, with a very short wavelength. Gamma rays are very penetrating, but do not cause as many ionisation events as alpha or beta radiation.

Geiger counter: This is an instrument based on a low-pressure gas-filled tube, which has a high voltage applied between the outer case and an inner electrode. The gas does not normally conduct, but if it is struck by ionising radiation, it will momentarily conduct electricity between the two electrodes, and this event is recorded by the Geiger counter. The frequency at which these pulses of current flow is proportional to the amount of ionising radiation the Geiger counter is exposed to. Many Geiger counters will indicate each pulse with a clicking sound – something which has become synonymous with radiation measurements.

Half-life: The time taken for something to decay or reduce in concentration to half of its original concentration. It is an easy way to quantify logarithmic decay. After one half-life has elapsed, the concentration will have reduced by half. After two half-lives, it will have reduced by half again (i.e., one quarter of the original amount). Half-life can be anything from tiny fractions of a second up to billions of years.

Generally, the shorter the half-life, the more dangerous a radioactive substance is.

Ingestion: A pathway for entering the human body via the mouth and swallowing.

ICBM: Inter Continental Ballistic Missile - a type of missile able to travel thousands of miles and deliver a weapons payload, which may be a nuclear or non-nuclear warhead.

Incandescent: When an object becomes so hot that it gives off energy in the form of visible light. The light it gives off is directly related to its temperature – from dull red through orange and yellow to bluish-white for extremely hot objects.

Ion: The name given to an atom when it becomes electrically charged by either receiving extra electrons or by having electrons taken away from it, so that the number of electrons no longer balances with the number of protons in its nucleus.

Ionization: The process of adding or removing electrons from atoms, to cause them to become electrically charged. This may be achieved with high temperatures, with high energy radiation, by bombardment with electrons, or by electrical discharges.

Ionizing radiation: Any form of radiation which is capable of stripping electrons from atoms, to produce ions. Ionising radiation can cause damage to human tissues and especially to DNA, which can in turn cause cancer.

Isotope: Any given element has a distinct number of protons in its nucleus. These are stuck together with neutrons. Most elements have a similar number of protons and neutrons in their atomic nucleus, but heavier elements tend to have a higher percentage of neutrons compared with protons. Within a given element, atoms may have fewer or more neutrons. Each one of these possible combinations is called an isotope. Some isotopes are stable, and some are less stable.

Kiloton (kt): As nuclear explosions are so much more powerful than conventional explosives, a convention has been adopted of measuring nuclear explosions by comparison to an equivalent amount of TNT chemical explosive. A 1kt nuclear explosion has the same energy equivalent to the explosion of 1000 tons of TNT. The smallest nuclear warhead ever made, the American W54 was rated at 0.1kt

Line of Sight: Where two radio stations are trying to communicate with each other, their signals may be absorbed and attenuated (reduced in intensity) by material that exists between them, such as by concrete, steel, animal or plant matter, topography or traffic. The best signal, or the best chance of making contact is when there is nothing lying in the path of the radio signal between the two radio sets – this is commonly referred to as 'line of sight'.

Logarithmic (or Exponential) Decay: A substance is said to decay exponentially or logarithmically if it decays at a rate that is proportional to its current amount. In simple terms, the amount decreases quickly to begin with, but appears to slow over time. It is most easily expressed in terms of half-life. The half-life is the time taken for half of an initial amount of radioactive material to decay. The half-life of radioactive decay is absolute, regardless of temperatures or pressure or how much material is present at the start - which allows us to predict very accurately what amount will remain in the future, if we know the half-life, and the amount present at the start.

Solid State Technology: Modern electronics based on semiconductors such as transistors and integrated circuits, are known as 'solid state'. This replaced older electronic technology based on vacuum tubes. The name 'solid state' comes from the fact that electricity passing through the electronics is passing through solid material, as opposed to passing through rarified gases inside vacuum tubes.

Megaton (Mt): A Megaton nuclear explosion is equivalent to a million tons of TNT (see kiloton, above).

Microbe: A microbe, also known as a microorganism, is a tiny living organism that is too small to be seen with the naked eye. For example, bacteria and viruses are types of microbes.

Molecule: A structure created when two or more atoms are chemically bonded together.

Mylar: Mylar is a brand name of DuPont Teijin Films™. It is a form of polyester, made from PET. However, the word 'Mylar' is often used indiscriminately to describe plastics which have a very thin coating of aluminium deposited on the surface, providing high barrier properties to massively reduce diffusion.

Neutron: The name given to sub-atomic particles within the nucleus of an atom, without an electric charge, which bind the protons together.

Non-ionizing radiation: Low energy radiation that does not cause atoms to become ionised and is therefore less harmful to living tissue. Electromagnetic radiation from radio waves, to microwave, infra-red and visible light are all non-ionising.

NORAD: North American Aerospace Defense Command – an early warning system, capable of detecting incoming nuclear missiles.

Nucleus: The name given to the large, central component of atoms, made up of neutrons and protons.

Penetrating radiation: The name for radiation that is capable of passing through the skin, potentially reaching and harming internal organs. Examples of penetrating radiation include high-energy beta particles, gamma rays, X-rays, and neutrons.

PPE ('Personal Protective Equipment'): This is the general name given for any equipment that is used to protect a person from exposure to hazards – for example, eye protection, face masks, breathing apparatus, protective gloves, protective clothing, ear defenders and so on.

Proton: One of the sub-atomic components of an atom, which is positively charged and is bound to other protons in the nucleus, by neutrons.

Radiation: The name for the transmission of energy from one point to another in the form of electromagnetic waves or particles.

Radiation sickness: The name given to the symptoms experienced by persons exposed to high levels of nuclear radiation.

Radioactivity: The emission of energy through changes in atomic nuclei, either due to natural causes, or to man-made nuclear processes. Such transformations may give rise to alpha, beta or gamma radiation or the emission of neutrons.

Radioisotope: An unstable isotope of an element, which is liable to decay, and in the process emit radioactivity of some kind. Some radioisotopes decay very quickly, others are more stable and can take millennia to decay. Some radioisotopes decay so slowly and have such long half-lives, that they have only very recently been classified as radioisotopes.

Radiological: Relating to ionizing radiation, such as nuclear radiation or X-rays.

Radionuclide: An unstable isotope of specific atomic mass, liable to radioactive decay.

Residence Time: In the Water Cycle, water passes through varies stages – such as rain, rivers, glaciers, lakes and sea. The average length of time a water molecule remains in a particular stage of the Water Cycle is known as the residence time for that particular stage. For example, water molecules have a residence time of approximately 9 days in the atmosphere, before falling as rain. At the other end of the scale, water in deep aquifers may remain there for 10,000 years before transforming to a different stage of the Water Cycle.

Semi-conductor: Modern electronics utilise materials which do not conduct as well as metals but conduct electricity better than non-metals. Most semi-conductors are based on the elements silicon or germanium, which have properties of both metals and non-metals. By adding in tiny amounts of other elements ('doping'), the conductive properties of such materials can be varied dramatically as voltage is applied. This is the basis of a transistor, which in turn forms the basis of integrated circuits ('chips'), and processors.

Shielding: The name given to any material placed between a radioactive source and a subject, to reduce the amount of radioactivity reaching the subject. Heavy dense materials are often the most suitable shielding – such as concrete or lead. To shield against neutrons, water is very effective, despite its relatively low density compared to other shielding materials.

Sievert (Sv): The international system (SI) unit for defining the equivalent absorbed dose of radiation. Some types of radiation are more biologically damaging than other types of radiation, and the Sv takes this into consideration.

Solid State Technology: Modern electronics based on semiconductors such as transistors and integrated circuits, are known as 'solid state'. This replaced older electronic technology based on vacuum tubes. The name 'solid state' comes from the fact that electricity passing through the electronics is passing through solid material, as opposed to passing through rarified gases inside vacuum tubes.

Sublimation: The process by which a chemical changes directly from a solid into a gas without passing through the intermediate liquid form. Sublimation can only occur under certain temperature and pressure conditions, which are specific to each chemical. Sublimation may occur when the temperature and pressure are below that of a chemical's 'triple point' (the combination of pressure and temperature at which a chemical may exist in any of the three forms of matter (solid, liquid or gas)

Surface burst: The name given for a nuclear explosion where the resultant fireball intercepts with the ground, causing surface material to be vaporised and sucked up into the mushroom cloud. Surface bursts create much more fallout, and the fallout tends to be highly radioactive.

Transceiver: This is the name given to a radio which can both transmit radio waves and receive radio waves, thus providing a two-way communication process.

Transmutation: In nuclear terms, transmutation occurs due to neutron capture, in which an atom either changes isotope by absorbing the neutron into the nucleus and thereby increasing the atomic mass by one unit ('isotropic transmutation') or if by absorbing a neutron, it becomes unstable and releases a beta particle, it does so by changing a neutron into a proton, which turns the atom into a different element, but with the same atomic mass. This is known as 'elemental transmutation'. Both types of transmutation are referred to as 'physical transmutation'

Tritium: This is an isotope of hydrogen in which the atomic nucleus contains two neutrons in addition to a proton. This form of hydrogen has triple the density of a normal hydrogen atom, and is radioactive, with a half-life of approximately 12 years. Because of its short half-life, it is almost non-existent in the natural environment. Tritium has been used to illuminate watches and also novelty key rings. Together with Deuterium, it is used as fuel for nuclear fusion reactors, and in thermonuclear bombs.

Volatile: A material is described as volatile if it is able to easily and quickly change from a liquid into a vapour form. Volatile chemicals will often evaporate away quickly if left exposed.

Thermonuclear Explosion: A nuclear bomb using conventional fission fuel which in turn creates a nuclear fission explosion where small nuclei such as hydrogen atoms are fused together into helium.

X-ray: A high-energy form of electromagnetic radiation, which can ionise atoms. X-rays do not carry as much energy as gamma rays, and therefore are not as harmful, but nevertheless are penetrating and harmful to human tissue.

When No One's Coming.

APPENDIX V: BIBLIOGRAPHY

Agency for Toxic Substances and Disease Registry (2002), *'Radon. Tox FAQs (TM).'*

Bilbrey J., Marrero C., Sassi M., Ritzmann A., Henson N. & Schram M., (2020), *'Tracking the Chemical Evolution of Iodine Species Using Recurrent Neural Networks.'*

Braverman E., Blum K., Loeffke B., Baker R., Kreuk F., Yang S. & Hurley J. (2014), *'Managing Terrorism or Accidental Nuclear Errors, Preparing for Iodine-131 Emergencies: A Comprehensive Review.'*

Centre for Disease Control and Prevention (2004), *'Feasibility Study of Weapons Test Fallout.'*

Centre for Disease Control and Prevention (c.2002) *'A Brochure for Physicians. ACUTE RADIATION SYNDROME.'*

Centre for Disease Control and Prevention (c.2002) *Brochure for Physicians. CUTANEOUS RADIATION INJURY.*

Centre for Disease Control and Prevention (2022), *'Radiation Emergencies.'*

Centre for Disease Control and Prevention (2018), *'Acute Radiation Syndrome (ARS): A Fact Sheet for the Public.'*

Nyffeler & Kaelin (2016); *'System Design and Assessment Notes. Note 47. EMP-Hardened Photovoltaic Generators: A Possible Emergency Power Solution for Critical Infrastructure.'*

Ellis VJ (1991), *'Effects of electromagnetic pulse (EMP) on cardiac pacemakers. Final report, Nov 88-Oct 89.'*

European Commission (2010), *'RADIATION PROTECTION NO 165. Medical effectiveness of iodine prophylaxis in a nuclear reactor emergency situation and overview of European practices. Final Report of Contract.'*, TREN/08/NUCL/SI2.520028.

Food Standards Agency (2017), *'The safety and shelf-life of vacuum and modified atmosphere packed chilled foods with respect to non-proteolytic Clostridium botulinum. Revision 3.'*

Landis J., Hamm N., Renshaw C., Dade W., Magilligan F & Gartner J. (2012), *'Surficial redistribution of fallout ^{131}iodine in a small temperate catchment.'*

Greenpeace (2008), *'Technical Brief: Problems with French European Pressurised Reactor at Flamanville.'*

Kheirabadi B.S., Scherer M.R., Estep J.S., Dubick M.A. & Holcomb J.B., (2009), 'Determination of efficacy of new hemostatic dressings in a model of extremity arterial hemorrhage in swine', in Trauma 67(3): 450-9.

Ofcom (2018), *'Citizens' Band (CB) radio spectrum use— information and operation. Of364. Guidance Publication.'*

Office for Nuclear Regulation (2013), *'Generic Design Assessment – New Civil Reactor Build'*

Raatjes G.J.& Smelt J.P., (1979), 'Clostridium botulinum can grow and form toxin at pH values lower than 4.6', in Nature 1979, Oct 4; 281 (5730):398-9.

Radiation Emergency Medical Management *(c.2020), 'Triage Guidelines Including Radiation Triage Guidelines.'*

Sharon, Halevy, Sattinger, Krantz, Admon, Banaim & Yaar (2014), *'Post Blast Nuclear Forensics of a Radiological Dispersion Device Scene.'*

Sumitani M., Takagi S., Tanamuray., Inque H, (2004), *'Oxygen Indicator Composed of an Organic/Inorganic Hybrid Compound of Methylene Blue, Reductant, Surfactant and Saponite in Analytical Sciences.'*

Thompson D.L., *(1976), Uranium in Dental Porcelain.*

U.S. DEPARTMENT OF HEALTH AND HUMAN SERVICES Agency for Toxic Substances and Disease Registry Division of Toxicology and Environmental Medicine (2002), *'Case studies in Environmental Medicine: RADIATION EXPOSURE FROM IODINE 131.'*

U.S. Department of Health and Human Services Food and Drug Administration Center for Drug Evaluation and Research (2001), *'Guidance Potassium Iodide as a Thyroid Blocking Agent in Radiation Emergencies.'*

U.S. Department of Health and Human Services Food and Drug Administration Center for Drug Evaluation and Research (2004), *'Guidance for Federal Agencies and State and Local Governments. Potassium Iodide Tablets Shelf-Life Extension.'*

U.S. Department of Health and Human Services Food and Drug Administration Center for Drug Evaluation and Research (2004), *'FDA Approves Drugs to Treat Internal Contamination from Radioactive Elements (Press Release).'*

U.S. Department of Health and Human Services Food and Drug Administration Center for Drug Evaluation and Research (2003), *'Guidance for Industry. Prussian Blue Drug Products— Submitting a New Drug Application.'*

United States Environmental Protection Agency (2012), *'A Citizen's Guide to Radon. The Guide to Protecting Yourself and Your Family from Radon.'*

When No One's Coming.

"Now I am become death, the destroyer of worlds."
...Robert Oppenheimer (Father of the atom bomb.)

Printed in Great Britain
by Amazon

90c6f938-456e-48e0-84b3-dac0d8e05805R01